MODERN FLUORESCENCE SPECTROSCOPY
4

MODERN ANALYTICAL CHEMISTRY

Series Editor: David Hercules
University of Pittsburgh

ANALYTICAL ATOMIC SPECTROSCOPY
By William G. Schrenk

PHOTOELECTRON AND AUGER SPECTROSCOPY
By Thomas A. Carlson

MODERN FLUORESCENCE SPECTROSCOPY, Volume 1
Edited by E. L. Wehry

MODERN FLUORESCENCE SPECTROSCOPY, Volume 2
Edited by E. L. Wehry

MODERN FLUORESCENCE SPECTROSCOPY, Volume 3
Edited by E. L. Wehry

MODERN FLUORESCENCE SPECTROSCOPY, Volume 4
Edited by E. L. Wehry

APPLIED ATOMIC SPECTROSCOPY, Volume 1
Edited by E. L. Grove

APPLIED ATOMIC SPECTROSCOPY, Volume 2
Edited by E. L. Grove

TRANSFORM TECHNIQUES IN CHEMISTRY
Edited by Peter R. Griffiths

ION-SELECTIVE ELECTRODES IN ANALYTICAL CHEMISTRY, Volume I
Edited by Henry Freiser

ION-SELECTIVE ELECTRODES IN ANALYTICAL CHEMISTRY, Volume 2
Edited by Henry Freiser

CHEMICAL DERIVATIZATION IN ANALYTICAL CHEMISTRY
Volume 1: Chromatography
Edited by R. W. Frei and J. F. Lawrence

MODERN
FLUORESCENCE
SPECTROSCOPY
4

Edited by
E. L. WEHRY
University of Tennessee
Knoxville, Tennessee

PLENUM PRESS · NEW YORK AND LONDON

Library of Congress Cataloging in Publication Data

Main entry under title:

Modern fluorescence spectroscopy.
  (Modern analytical chemistry)
  Includes bibliographical references and indexes.
  1. Fluorescence spectroscopy. I. Wehry, E. L., 1941 –   . [DNLM: 1. Spec-
trometry, Fluorescence. QD 96.F56 M689]
QD96.F56M6   543'.0858   75-43827
ISBN 0-306-40691-8   AACR1

© 1981 Plenum Press, New York
A Division of Plenum Publishing Corporation
233 Spring Street, New York, N.Y. 10013

Printed in the United States of America

Contributors

James B. Callis, Department of Chemistry, University of Washington, Seattle, Washington 98195

Gary D. Christian, Department of Chemistry, University of Washington, Seattle, Washington 98195

Ernest R. Davidson, Department of Chemistry, University of Washington, Seattle, Washington 98195

DeLyle Eastwood, Chemistry Branch, U.S. Coast Guard Research and Development Center, Avery Point, Groton, Connecticut 06340

G. M. Hieftje, Department of Chemistry, Indiana University, Bloomington, Indiana 47405

Gleb Mamantov, Department of Chemistry, University of Tennessee, Knoxville, Tennessee 37916

Jeffery H. Richardson, General Chemistry Division, Lawrence Livermore Laboratory, University of California, Livermore, California 94550

T. Vo-Dinh, Health and Safety Research Division, Oak Ridge National Laboratory, Oak Ridge, Tennessee 37830

E. E. Vogelstein, Department of Chemistry, Indiana University, Bloomington, Indiana 47405

E. L. Wehry, Department of Chemistry, University of Tennessee, Knoxville, Tennessee 37916

John C. Wright, Department of Chemistry, University of Wisconsin, Madison, Wisconsin 53706

Preface

Since the appearance of the first two volumes of *Modern Fluorescence Spectroscopy* in 1976, important advances continue to be made in both the techniques and applications of molecular luminence. In terms of "hardware," it is only recently that the application of laser excitation to molecular fluorometry has become feasible under conditions that are analytically realistic. The improvements that can be effected in sensitivity, analytical selectivity, and ability to handle "difficult" samples by laser fluorometry have only begun to be exploited. Likewise, time-resolved fluorometry has received widespread use in fundamental studies (a sizable number of which deal with biological systems), but has as of yet received relatively little analytical utilization.

The use of electronic array detectors offers the promise of obtaining luminescence spectra more rapidly, and perhaps ultimately with greater sensitivity, than is possible by the use of scanning instruments equipped with conventional detectors. The increasing capabilities of microcomputers and the increasing sophistication of "smart" spectroscopic instrumentation signify that much more efficient acquisition and use can now be achieved of the information contained in the "excitation–emission matrix" inherent in the luminescence phenomenon.

Increasingly elegant applications are being reported of these (and other) relatively new techniques in both the biological sciences and analytical chemistry. The design and interpretation of "fluorescent probe" experiments in biological systems continue to increase in sophistication and specificity. Newly developed capabilities for rapid acquisition of luminescence spectra of very small samples (exemplified by studies of the transport and reaction kinetics of fluorescent species in single cells) have important implications for the life sciences. Laser excitation promises to have increasingly widespread impact for users of fluorescence spectrometry in both fundamental and analytical investigations in biological systems.

Probably the most important aspect of modern fluorescence spectroscopy to the analytical chemist is the greatly enhanced capability to deal with multicomponent samples. Increased analytical selectivity is now being attained by a variety of routes: more specific chemical techniques (e.g., immunofluorescence); the combination of luminescence with chromatographic separations; use of temporal, as well as spectral, resolution of components; the use of laser excitation, coupled with special sampling techniques, to achieve very high spectral resolution under analytically realistic conditions; and more effective use of the information content of excitation and emission spectra.

It is the purpose of Volumes 3 and 4 of this series to survey some of the more important recent developments in these areas. As in Volumes 1 and 2, we have no intention of providing a "comprehensive" examination of the field nor to develop material treated in the "standard texts." Each of the individual chapters is self-contained. The individual authors have been encouraged not only to develop the current status of their respective topics, but also to include a modicum of informed speculation of future trends in each area. It is our hope that these volumes will assist in overcoming some of the communication barriers between the "developers" and the "consumers" of new fluorometric techniques.

It is a distinct pleasure for me to acknowledge, with gratitude, the contributions of the individual chapter authors. The excellence of their scientific work and their willingness to communicate both the technical details and the broader motivation and implications of their research have made these volumes possible.

Knoxville, Tennessee   E. L. Wehry

Contents

3. Probe Ion Techniques for Trace Analysis
John C. Wright

4. Array Detectors and Excitation–Emission Matrices in Multicomponent Analysis
Gary D. Christian, James B. Callis, and Ernest R. Davidson

5. Synchronous Excitation Spectroscopy

T. Vo-Dinh

6. Low-Temperature Fluorometric Techniques and Their Application to Analytical Chemistry
E. L. Wehry and Gleb Mamantov

7. Use of Luminescence Spectroscopy in Oil Identification
DeLyle Eastwood

Contents of Other Volumes

VOLUME 3

Chapter 1

Applications of Lasers in Analytical Molecular Fluorescence Spectroscopy

Jeffery H. Richardson

A. INTRODUCTION

Since their inception lasers have been described as a solution in search of a problem. It was recognized quite early that lasers would be an ideal excitation source for spectroscopic analysis. The actual application of lasers to spectroscopy has, however, proceeded in a rather sporadic fashion. For example, laser technology completely rejuvenated Raman spectroscopy. In the space of a few short years, virtually all conventional Raman spectroscopy is done with lasers,[1] and a host of nonlinear Raman techniques have sprung up that capitalize on the unique spectral properties of lasers.[2] On the other hand, elemental spectroscopic analysis is still largely performed by conventional nonlaser techniques.[3]

The application of lasers to specific analytical techniques has closely paralleled the development of lasers. Conventional Raman spectroscopy requires a stable, reliable single-frequency laser, which has been available for some time in the form of rare-gas-ion lasers. Elemental spectroscopic analysis requires in addition that the laser source be tunable (or have many fixed frequencies); laser technology is just beginning to develop widely tunable, stable, and reliable sources.

JEFFERY H. RICHARDSON • General Chemistry Division, Lawrence Livermore Laboratory, University of California, Livermore, California 94550

The current state of laser technology with respect to analytical molecular fluorescence spectroscopy lies somewhere in between that of Raman and elemental analysis. The requirement for tunability is not so severe. However, additional characteristics of fluorescence (especially the relatively short lifetimes) impose additional requirements on the laser excitation source. The ideal laser excitation source for molecular fluorescence is not yet available. Even when this ideal excitation source is available, however, the ultimate in analytical molecular fluorescence spectroscopy will not be achieved. The problems in current analytical molecular fluorescence spectroscopy lie not only in the characteristics of the excitation source, but also in the very nature of the fundamental fluorescence process. Selectivity of the fundamental process will still remain a problem. The matrix frequently determines not only the extent of selectivity, but also the limit of detection, so achieving the ultimate in sensitivity will also remain a problem. In the near future neither of these problems appears to have laser solutions.

It is the purpose of this chapter to illustrate how lasers and laser technology can be applied to the problems of analytical molecular fluorescence spectroscopy. The use of lasers as excitation sources in analytical fluorescence has led to a significant improvement, compared to conventional excitation sources, in the minimum detectability for specific compounds. Furthermore, the use of laser techniques has resulted in new analytical processes that were not possible with conventional sources. Finally, the use of lasers in conjunction with other techniques and technologies offers the promise of greatly expanding the types of problems for which analytical molecular fluorescence spectroscopy can be a solution.

B. INSTRUMENTATION

1. Lasers

There are three salient properties that distinguish lasers from conventional light sources: spatial coherence, degree of monochromaticity, and temporal pulsewidth. Other properties, such as polarization and temporal coherence, are less significant in analytical applications.

Lasers can be readily focused down to spot sizes on the order of microns ($\lambda/4$ is the diffraction limit). This property is most advantageous in obtaining spatial information on the composition of surfaces (e.g., laser Raman microprobe[4]). Other applications include focusing onto chromatography effluents[5] and interrogating a sample spatially with two crossed laser beams. Remote atmospheric monitoring also capitalizes on the spatial coherence of laser beams.

Lasers can be extremely monochromatic, down to the order of Hz,[6] although typical bandwidths of commercial instruments are on the order of several megahertz. Wavelength resolution is not a problem when using lasers as an excitation source for solution analysis, although excitation of gases (e.g., OH,[7] NO_2[8]), glasses,[9] or crystals[10] does take advantage of the laser's high monochromaticity and spectral brightness (high intensity per unit bandwidth).

Tunability is a more important consideration. Many lasers have only a few fixed frequencies (e.g., rare-gas-ion lasers, rare-gas–halogen excimer lasers, He–Cd, N_2, Nd). Tunability in the visible region is commonly achieved with dye lasers, although the tunable range for any single dye is limited (Figure 1).[11,12] Generation of tunable fundamental radiation extends from the near infrared (1200 nm[13]) into the near ultraviolet (320 nm, p-terphenyl). To go further into the ultraviolet, which is the region of greatest interest for fluorescence excitation, requires nonlinear techniques. Either pulsed or cw sources can be doubled to about 258 nm with modest efficiencies using KDP (potassium dihydrogen phosphate) crystals. Further extension into the ultraviolet requires processes of lower efficiencies or higher thresholds [potassium pentaborate (KB 5), stimulated Raman scattering[14]]. These processes require higher peak powers; thus the deep UV is almost solely the domain of pulsed lasers.

The biggest single advantage of lasers over conventional sources as a fluorescence excitation source lies in their ability to produce a short temporal pulse of nearly monochromatic light. Pulse widths vary from microseconds (flashlamp-pumped lasers) through nanoseconds (excimer and nitrogen lasers) to picoseconds and beyond (mode-locked lasers). Narrow pulse widths offer at least two distinct advantages: (1) Background scatter and perhaps interfering fluorescence can be temporally reduced; (2) pulse widths no longer need to be deconvolved from fluorescence decay curves. Thus mixtures can sometimes be resolved into their components on the basis of the various fluorescence lifetimes. Time-correlated single-photon-counting techniques, now limited mainly by jitter in the photomultiplier tube, can be used to measure short fluorescence lifetimes[15] (although a new technique permits this to be done with cw lasers[16]). Of course, for very short lifetimes the photomultiplier tube response becomes the limiting factor and must be deconvoluted.[17,18]

There are several different ways to produce short laser pulses with commercially available instrumentation. High-peak-power lasers (nitrogen, excimer, or neodymium) intrinsically have pulse widths between 5 and 40 nsec, with repetition rates of a few hertz. Such pulse widths have some

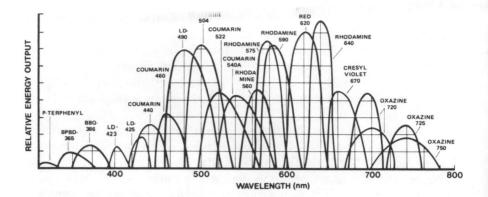

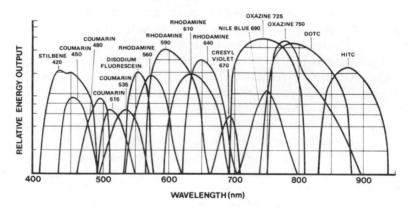

FIGURE 1. Relative tuning curves for various dyes with commercial lasers (courtesy of Exciton Chemical Co., Dayton, Ohio). (Top) Pulsed coaxial flashlamp-pumped dye laser (courtesy of Phase-R Corporation, New Durham, New York); (bottom) cw dye laser pumped with an argon-ion or krypton-ion laser (courtesy of Spectra Physics, Inc., Mountain View, California).

applicability in fluorescence analysis of long-lived fluorophors[19] (e.g., polycyclic aromatic hydrocarbons, for which the fluorescence lifetimes can be as long as several hundred nanoseconds).

Argon-ion or krypton-ion gas lasers, nominally cw lasers, can be turned into pulsed lasers by cavity dumping.[20] This technique involves an intracavity acousto-optic deflector, with the net result that a variable pulse width (10 nsec to cw) at a variable repetition rate (several MHz to single shot) can be obtained (the repetition rate and pulse width are, of course, interrelated). Unlike nitrogen or excimer lasers, cavity-dumped ion lasers are low-peak-power (≤ 50 W) low energy excitation sources, but with modest average powers (up to ~ 1 W).

Subnanosecond pulses are required for most fluorescence lifetime measurements. The generation of such pulses requires mode-locking techniques.[11,21] This can be done passively (saturable absorber)[22–24] or actively (acousto-optic loss modulation).[25] Actively mode-locked ion lasers can be used to synchronously pump a dye laser,[26,27] resulting in tunable picosecond pulse widths at a fixed megahertz rate (determined by the laser cavity length; Figure 2). The repetition rate can be reduced either by cavity dumping or pulse picking external to the cavity.[28,29] Such instrumentation is now commercially available. Additional repetition of these techniques can result in subpicosecond pulses.[30]

It is important to remember that all laser characteristics cannot be obtained simultaneously. Obviously subpicosecond pulses will have relatively broad linewidths; it is a good laser system that is Fourier-transform limited. Only recently have tunable picosecond sources been available in the blue[27,31] as well as in the orange portion of the spectrum. Megahertz repetition-rate lasers, so useful for rapid data acquisition and signal averaging, have low energy pulses with modest peak powers. Consequently, such systems are not efficiently frequency shifted to the deep ultraviolet (≤ 258 nm) by the various nonlinear techniques. Most nonlinear techniques work best with high-peak-power lasers that invariably have a low repetition rate and nanosecond pulse widths. A picosecond pulse width with high peak power has been generated by using a neodymium-pumped amplifier with a mode-locked cavity-dumped dye laser[30]; the output pulse, less than one picosecond wide and more than a gigawatt, is sufficiently intense to generate a continuum from water. This continuum can be used for transient absorption studies.

The emphasis for future laser development is largely placed on developing better ultraviolet lasers. A prominent candidate is the excimer laser[32]; already commercial instruments can deliver many megawatts at wavelengths as short as 193 nm (ArF). The further development of dyes is required before an excimer-pumped dye laser can be used to generate tunable radiation throughout the ultraviolet spectrum. Metal-vapor lasers are currently being investigated as cw laser sources for the ultraviolet.[33] Tunable laser sources have already been integrated with sophisticated spectrofluorometer systems,[34] and this trend will continue as laser sources become more flexible, reliable, and applicable to various real life problems.

Jeffery H. Richardson

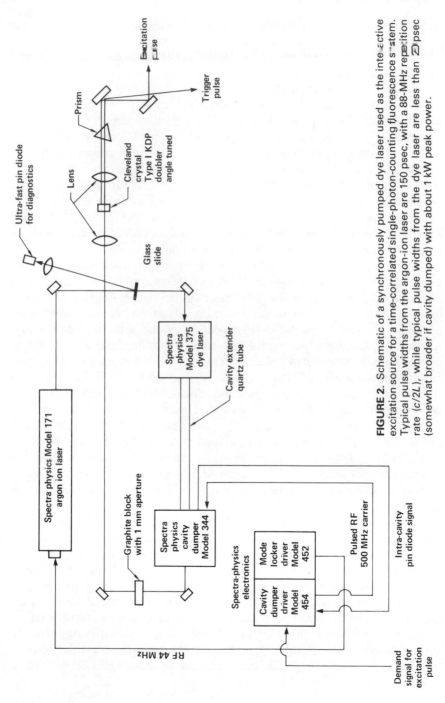

FIGURE 2. Schematic of a synchronously pumped dye laser used as the interactive excitation source for a time-correlated single-photon-counting fluorescence system. Typical pulse widths from the argon-ion laser are 150 psec, with a 88-MHz repetition rate (c/2L), while typical pulse widths from the dye laser are less than 20 psec (somewhat broader if cavity dumped) with about 1 kW peak power.

2. Detection Systems

Detection systems for fluorescence measurements have been previously described.[35] With respect to laser applications, both cost and the choice of the laser excitation system combine to determine the optimal detection system. The lowest limits of detection have invariably been achieved with pulsed lasers. The duty cycle of pulsed lasers is invariably so low (e.g., 10^{-7} for a N_2 laser at 10 Hz) as to preclude detection systems employing lock-in amplifiers. Consequently, gated detection systems become the method of choice, employing either boxcar integration or photon counting.

Boxcar integration[36] is the simplest, least expensive, and most convenient to use. Sub-part-per-trillion limits of detection have been achieved with boxcar integration and either a low-repetition-rate high-peak-power laser or a high-repetition-rate low-peak-power laser.[37] Figure 3 illustrates a complete analytical fluorescence instrument employing boxcar detection with a low-repetition-rate high-peak-power laser. Temporal resolution on the order of 10 nsec, sufficient for many applications, can be obtained with up to 5-MHz repetition rates (as would be available from a cavity-dumped

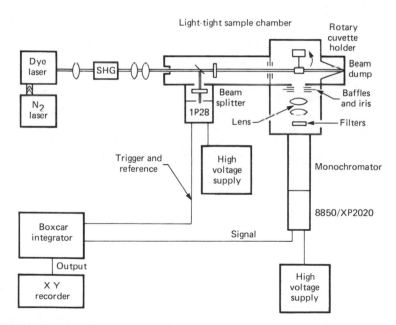

FIGURE 3. Schematic of a fluorescence system employing a high-peak-power low-repetition-laser and boxcar detection. From Richardson and Ando,[19] reproduced by permission of the American Chemical Society.

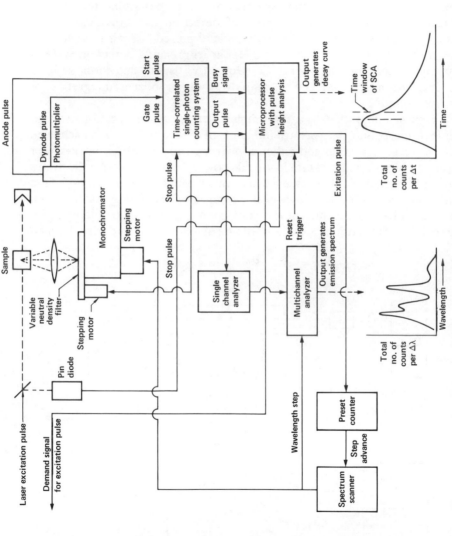

FIGURE 4. Schematic of the instrumentation necessary for an interactive time-correlated single-photon-counting fluorescence system. The microprocessor requests an excitation pulse from a cavity-dumped synchronously pumped dye laser after a fluorescence pulse has been detected.

laser system). The maximum temporal resolution is obtained with a sampling unit (approximately 100 psec, limited by both the sampling head rise time and main-frame jitter), with a corresponding data acquisition rate of 50 kHz.

Photon-counting techniques are less suited to high-peak-power low-repetition-rate lasers. Lower light levels are obtained with high-repetition-rate low-peak-power lasers, but even there a simple ratemeter was found to be inferior to boxcar detection.[37] Time resolution invariably results in lower detection limits with pulsed excitation sources by eliminating scatter and short-lived background fluorescence.[19] Consequently, time-correlated photon-counting techniques are needed to obtain limits of detection comparable to those obtained with boxcar detection.[38] Figure 4 illustrates a state-of-the-art time-correlated single-photon-counting system using a synchronously pumped dye laser excitation source (i.e., high repetition rate, low peak power).[38,39] Temporal resolution is on the order of 150 psec, limited by the transit time spread in the photomultiplier tube, with maximum data acquisition rates on the order of 150 kHz. The system shown in Figure 4 was used to quantitatively determine rubrene in benzene, with a limit of detection equal to 5 parts per trillion.[38] The analytical working

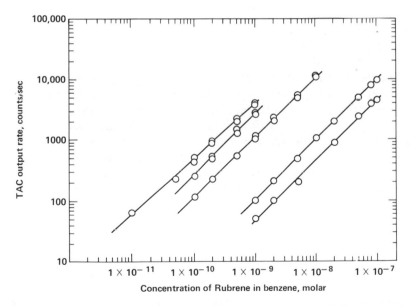

FIGURE 5. Analytical curves for fluorescence detection of rubrene in benzene using a synchronously pumped dye laser excitation source and interactive time-correlated single-photon-counting techniques.

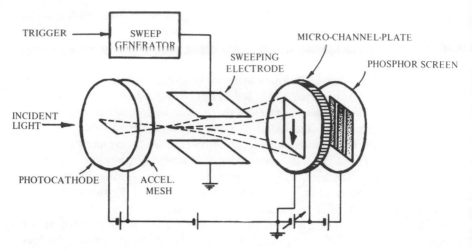

FIGURE 6. Schematic illustrating the principles of operation of a streak camera. Incident light is focused onto the photocathode of the streak tube. The emitted electrons are accelerated into the deflection field, which is synchronized with the arrival of the electron field and ramped with time. The net effect is to sweep the electrons vertically in time, resulting in an image on the microchannel plate that has time as the vertical axis (courtesy of Hamamatsu Corporation, Middlesex, New Jersey).

curves were linear over four orders of magnitude (Figure 5). The disadvantages of time-correlated single-photon-counting detection in an analytical application are primarily the longer data acquisition times (100 sec for the rubrene data versus an approximately 2-sec time constant for boxcar detection), the greater complexity, and the greater expense.

The ultimate in temporal resolution is currently obtained with a streak camera. Figure 6 is a schematic illustration of the principles of operation of a commercially available streak camera. Maximum resolution is on the order of 1 psec for a single sweep, although repetitive sweeps experience several psec time jitter. Repetition rates vary between commercial instruments from kilohertz to megahertz. Figure 7 illustrates how a streak camera might be employed to obtain time-resolved fluorescence or emission spectra from transient species.[40] Signal integration can be accomplished with the intensified silicon intensified target (ISIT) in the streak camera, the ISIT in the optical multichannel analyzer (OMA), and/or digitally after the signal has been read out and stored. The use of streak cameras for temporal resolution eliminates the need for deconvolution of the photomultiplier tube response. Also, streak cameras are much more flexible for measuring mode-locked laser pulse widths than doubling or two-photon fluorescence techniques because streak cameras have less wavelength restriction.

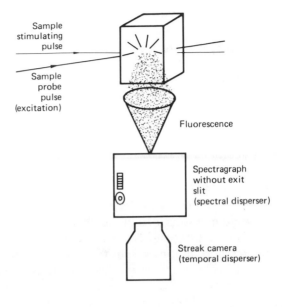

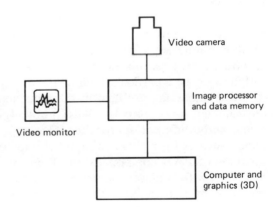

FIGURE 7. Block diagram for instrumentation required for fast transient analysis. The streak camera output has time as the vertical scale and wavelength as the horizontal scale; the trigger for the streak camera can be provided by either the stimulating or the probe pulse.

Streak cameras have been primarily used to obtain fluorescence lifetime data as a function of the local environment[41-43] or for fast kinetics.[44] A mode-locked laser has been used as the excitation source for these investigations. The application of streak cameras in analytical molecular fluorescence spectroscopy is still in the future, but it can be anticipated that they will provide hitherto unavailable diagnostic information on transient phenomena in flash photolysis, shock waves, explosions, and combustion.

C. APPLICATIONS

1. Polycyclic Aromatic Hydrocarbons (PAH)

The detection of PAH compounds has assumed increased importance recently because of environmental considerations. Many PAH compounds are potent carcinogens,[45] and an increasing dependence on fossil fuels will result in the greater release of PAH compounds into the environment.

Detection of PAH compounds is a question of both sensitivity and selectivity. Their detection has been successfully attempted with fluorescence techniques, partly because fluorescence is so sensitive and partly because PAH compounds constitute almost an ideal class of fluorophors.[46] PAH compounds have large extinction coefficients ($\geq 10^4 M^{-1}$ cm^{-1}), with maximum absorptions in the near ultraviolet or visible, large fluorescence quantum yields (usually ≥ 0.2 and in many cases approaching 1.0), and long fluorescence lifetimes (tens to hundreds of nanoseconds).

Table I summarizes the minimum detectabilities of several PAH compounds in water obtained with the experimental apparatus depicted in Figure 3.[19] These limits of detection are generally two orders of magnitude lower than those obtained using conventional sources.[47] A linear dependence of fluorescence intensity on PAH concentration was obtained, sometimes extending over six orders of magnitude. In many cases the temporal resolution inherent in pulsed laser techniques was used to eliminate background scatter and fluorescence. This technique can be used even in very turbid samples; a particularly dramatic example is depicted in Figure 8.[48]

The remaining problem in the detection of PAH compounds in the environment is selectivity. This problem is being addressed by two significantly different approaches. One approach is to use another technique, such as high-pressure liquid chromatography (HPLC), to separate mixtures of PAH compounds and then use fluorescence detection. This technique with nonlaser sources has resulted in minimum detectabilities on the order of picograms.[49] Current findings with laser excitation result in similar sensitivities.[50] However, once again one advantage of using laser excitation is the use of temporal resolution, which offers another selectivity parameter in addition to excitation wavelength, emission wavelength, and retention time (Figure 9). The coupling of the high sensitivity of laser-induced fluorescence techniques with the selectivity of chromatographic techniques has been used to great advantage in the analysis of aflatoxins.[50-52]

The second approach is to sufficiently enhance the wavelength selectivity inherent in fluorescence analysis so that complex mixtures of PAH compounds can be separated. Enhancement of wavelength selectivity in

Table 1. Limits of Detection of Some Polycyclic Aromatic Hydrocarbons (PAH) by Laser-Induced Molecular Fluorescence

PAH	Transition of interest[a]	Excitation wavelength, λ_{ex} (nm)	Absorptivity at λ_{ex} (M^{-1} cm^{-1})	Fluorescence wavelength, λ_{em} (nm)	Limits of Detection	
					M	μg/liter (ppb)
Benzene	$^1A \to {}^1L_b$	259.95	9×10^1 [b]	302, 273	2.5×10^{-7}	19
Naphthalene	$^1A \to {}^1L_a$	273.00	6×10^3 [c]	340, 360	1×10^{-11}	1.3×10^{-3}
Anthracene	$^1A \to {}^1B_b$	258.70	5×10^3 [c]	404	5×10^{-11}	8.9×10^{-3}
		254.00	1×10^5 [c]		$<2.5 \times 10^{-11}$	$<4.4 \times 10^{-3}$
Fluoranthene	$^1A \to {}^1B_b$	287.00	4×10^4 [b]	450	5×10^{-12}	1×10^{-3}
Pyrene	$^1A \to {}^1B_b$	273.00	6×10^4 [c]	395	2.5×10^{-12}	0.5×10^{-3}

[a] Reference 47.
[b] Methanol/water.
[c] Methanol.

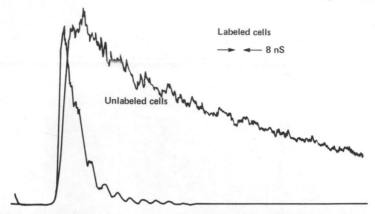

FIGURE 8. Time-resolved laser fluorescence signals of unlabeled mouse cells and mouse cells treated with benzo[a]pyrene (and hence with diol-epoxide benzo[a]-pyrene bound to DNA). The excitation frequency was 316 nm (frequency-doubled output of a nitrogen-pumped dye laser using cresyl violet) and the emission was monitored at 400 nm. Although the solution was very turbid, and both kinds of cells fluoresce and scatter around 400 nm, the much longer fluorescence lifetime of the labeled cells makes it possible to detect them unambiguously. From Richardson *et al.*,[48] reproduced by permission of the U.S. Government Printing Office.

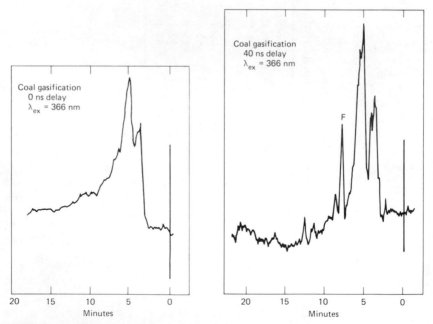

FIGURE 9. Liquid chromatogram of real life sample taken from the condensed steam distillate obtained during a coal-gasification burn near Gillette, Wyoming (Hoe Creek No. 3). Fluoranthene (F), unrevealed at zero time delay, is clearly indicated at 40 nsec delay.

fluorescence analysis can be enhanced by matrix isolation[53,54] or Shpol'skii techniques.[55] The use of narrow-band laser excitation with matrix-isolated[56] or frozen glass[57] samples of PAH compounds further enhances the selectivity (Figure 10). Finally, the use of time-resolved laser-induced fluorescence and line-narrowing matrix techniques enhances still further the ability to perform sensitive and selective analyses of complex PAH mixtures.[58] Real life samples will always present a greater challenge, but great progress in the sensitive and selective analysis of PAH compounds in the environment has been made.

Several of these approaches are discussed in more detail in Chapters 6 and 7 of the present volume.   *197*

2. Biochemicals

A great deal of biochemical analysis is done with fluorescence techniques. Not only are intrinsic fluorophors assayed, but extensive use of fluorescent labeling and staining techniques is employed to enable nominal nonfluorophors to be analyzed by fluorescence techniques. In many cases the sensitivity of these techniques has been enhanced by the use of laser-induced fluorescence.

Table II summarizes data for several biochemicals that were obtained with the apparatus illustrated in Figure 3.[59–61] The typical improvement in the limit of detection was two orders of magnitude. In two cases, the photolabile vitamins A and B_6, significant sample decomposition occurred with the intense levels of radiation present in a nitrogen laser pulse. With boxcar detection, however, the total irradiation time could be kept reasonably short.

Biochemical fluorescence analysis can frequently take advantage of the sensitivity of the fluorescence lifetime to the local environment.[42,43,62] For example, the lifetime of ethidium bromide dramatically increases upon intercalation into DNA. The lifetimes of several fluorescamine-labeled amino acids were measured with a synchronously pumped dye laser, illustrating the advantages of such a laser system for biochemical analysis.[29] The label lifetime was observed to vary as a modest function of the peptide length. Also, it is in biochemical analysis that fluorescence polarization techniques find their greatest application (e.g., immunofluorescence studies[63]). Consequently, in such studies the polarization properties of laser excitation could be exploited to their fullest.

The need also exists for extremely sensitive and selective analysis of potentially toxic or carcinogenic chemicals in food analysis. Aflatoxin analysis provides an excellent example of the application of lasers to analytical molecular fluorescence spectroscopy.[5,51,52] The experimental

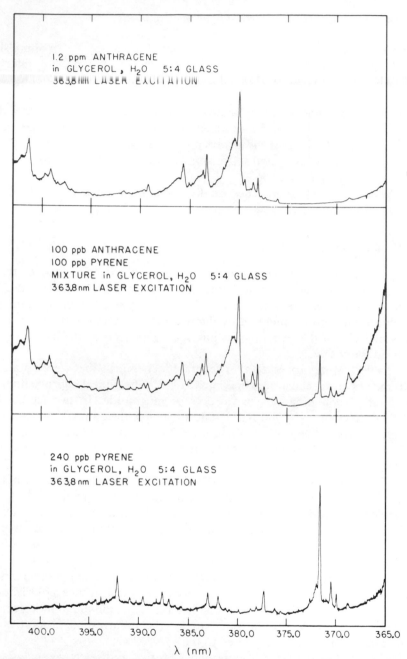

1.2 ppm ANTHRACENE
in GLYCEROL, H_2O 5:4 GLASS
363.8 nm LASER EXCITATION

100 ppb ANTHRACENE
100 ppb PYRENE
MIXTURE in GLYCEROL, H_2O 5:4 GLASS
363.8 nm LASER EXCITATION

240 ppb PYRENE
in GLYCEROL, H_2O 5:4 GLASS
363.8 nm LASER EXCITATION

λ (nm)

FIGURE 10. Resolution of a mixture of two PAH compounds (anthracene and pyrene) in a frozen glass by fluorescence line narrowing. The excitation wavelength was 363.8 nm and the temperature was 4.2°K. From Brown et al.,[57] reproduced by permission of the American Chemical Society.

TABLE II. Limits of Detection of Some Biochemicals by Laser-Induced Molecular Fluorescence

Compound	Excitation wavelength, λ_{ex} (nm)	Absorptivity at λ_{ex} (M^{-1} cm^{-1})	Fluorenscence wavelength, λ_{em} (nm)	Limits of detection	
				M	g/l (ppb)
Vitamin A acetate	337.1	1.8×10^4	500	3×10^{-9}	1.0
Vitamin B$_2$ (riboflavin)	375.0	1.1×10^4	540	1.25×10^{-12}	0.47×10^{-3}
FAD	267.0	3.8×10^4	525	5×10^{-11}	39×10^{-3}
Vitamin B$_6$ (pyridoxine)	337.1	1.6×10^3	410	2.5×10^{-10}	50×10^{-3}
Vitamin B$_{12}$ (cyanocobalamin)	260.0	1.2×10^4	310	ca.1×10^{-7}	1.4×10^2
Tryptophan	270.0	5.4×10^3	358	2.5×10^{-10}	50×10^{-3}
Aniline/fluram	390.0	1.7×10^4	500	3.8×10^{-11}	3.5×10^{-3}
Arginine/fluram	390.0	$\sim 1.7 \times 10^4$	500	5×10^{-11}	10×10^{-3}
Tyrosine/fluram[a]	390.0	$\sim 1.7 \times 10^4$	500	3.8×10^{-11}	7×10^{-3}
Tyroxine/fluram[a]	390.0	$\sim 1.7 \times 10^4$	500	5×10^{-11}	40×10^{-3}

[a] J. H. Richardson, unpublished data.

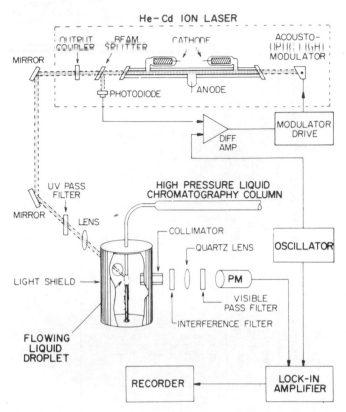

FIGURE 11. Schematic of the laser-induced fluorescence, high-pressure liquid chromatography instrumentation used to detect aflatoxins in corn. The feedback-stabilized laser is 100% modulated at 50 kHz and excites fluorescence at 325 nm. From Diebold and Zare,[52] reproduced by permission of the American Chemical Society.

instrumentation used to detect aflatoxins is schematically depicted in Figure 11. This instrumentation combines the sensitivity of laser-induced fluorescence and the selectivity of high-pressure liquid chromatography. Subpicograms of aflatoxin standards were detected (Figure 12); the minimum amount of material actually detected in the chromatographic effluent was less than 7 fg. Real-life samples were also analyzed with this instrumentation; 7-ppb aflatoxin B_1 was detected in a corn sample.

3. Dyes

The earliest examples of the extreme sensitivity achievable with lasers in analytical fluorescence spectroscopy came from dye analysis (aflatoxins

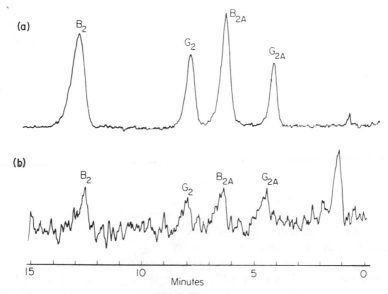

FIGURE 12. (a) Chromatogram of aflatoxins B_1, G_1, B_2, and G_2 eluting from a C18 μBondapak HPLC column, 30 pg each, $\lambda_{ex} = 325$ nm. The aflatoxins B_1 and G_1 have been converted to aflatoxins B_{2A} and G_{2A}, respectively, with HCl. (b) The fluorometer response at the detection limit of 750 fg with a 3-sec time constant. From Diebold and Zare,[52] reproduced by permission of the American Chemical Society.

are chemically similar to coumarins, a class of laser dyes). Rhodamine 6G, a common laser dye, has been detected at a concentration level of 0.6 parts per trillion; the analytical working curve has a linear dependence on concentration for over six orders of magnitude.[64] Other examples of sensitive dye detection with laser-induced fluorescence include rhodamine B,[37] fluorescein,[65] and fluorescein as a tag for immunoassay.[66,67] Winefordner has listed many of the compounds that have been analyzed by laser induced fluorescence.[68]

Currently the practical limit of detection by laser-induced fluorescence in solution appears to be approximately 1 part per trillion ($10^{-12}\,M$) for ideal fluorophors, even though the theoretical limit is much lower.[69] This limit is largely dictated by background considerations; it has been noted that the residual fluorescence for the better solvents corresponds roughly to a concentration–quantum yield product of 10^{-12}.[70]

4. Unique Laser Techniques

The most unique feature of laser sources is the high spectral power that can be achieved. The power is sufficient to produce multiphoton effects; for example, two-photon fluorescence is a common way of measuring narrow temporal pulse widths. Two counter propagating pulses are made to overlap in time and space in a dye solution; fluorescence results only while the two pulses simultaneously excite the dye.

The theoretical[71] and experimental[72] considerations in two-photon fluorescence have been discussed. The analytical applications of two-photon fluorescence have been few; a notable exception is that of two-photon excited molecular fluorescence in optically dense media.[73] Sub-micromolar detection limits were achieved with signal intensity proportional to the square of the power. A significant advantage is the ability to bulk excite the sample, rather than just performing front-surface fluorescence.

One potential advantage of using a multiphoton excitation scheme would be to enhance the selectivity of molecular fluorescence. The use of two different lasers, and hence two different wavelengths, greatly increases the possibilities of enhancing selectivity (e.g., atomic uranium isotope separation). Even more selectivity can be obtained by using an infrared laser and a visible laser. The infrared laser can distinguish compounds to the extent that their infrared absorption spectra are distinct. The visible laser provides sensitive, albeit not very selective, electronic excitation, with subsequent fluorescence or photoionization detection. One notable example of this concept is the uranium isotope separation process based on UF_6. There are few examples of two-photon visible–infrared processes in condensed phases, owing to the rapid vibrational relaxation rates. However, Figure 13 illustrates an infrared selective visible fluorescence signal from fluorophors in a condensed phase.[74] The decrease in fluorescence intensity is attributed to a selective thermal effect, increasing the rate of nonradiative decay when the infrared laser radiation is absorbed. It is also possible that laser thermal lens effects,[75] with fluorescence detection and selective infrared excitation, would also lead to selectivity enhancement in laser-

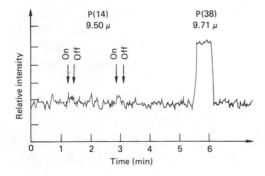

FIGURE 13. Representative spectrum illustrating the selectivity of two-photon infrared–visible laser-induced fluorescence. A solid sample of rubrene was irradiated with a chopped CO_2 laser at either 9.50 μm (P14) or 9.71 μm (P38), with simultaneous monitoring of fluorescence excited by a coincidence argion-ion laser at 488 nm. Rubrene has appreciable infrared absorption at 9.71 μm, but not at 9.50 μm.

induced fluorescence.[76] As laser techniques become more common, additional effects based on the simultaneous use of two or more lasers will result in new applications to analytical molecular fluorescence.

The spectral intensity and spatial coherence of lasers is utilized in remote monitoring, most notably of atmospheric pollutants.[77] However, the advent of fiber optics has made it possible to utilize laser fluorescence techniques for remote, non-line-of-sight monitoring. For example, two optical fibers, one transmitting a laser excitation pulse and one the resulting fluorescence, have been used to detect laser dyes in solution at the parts per-billion level.[78] Ultimately, one master laser fluorescence system can be envisioned that monitors fluorescence intensity at several remote sites (analogous to a master computer with remote terminals).

E. CONCLUSIONS

Lasers have already made an impact in analytical molecular fluorescence spectroscopy,[79] primarily by enhancing the sensitivity of fluorescence analysis. This enhancement in sensitivity over conventional sources is primarily due to the higher spectral powers and narrower temporal bandwidths obtainable with lasers. As lasers become more reliable and more widely tunable they will continue to supplant conventional sources wherever high sensitivity is needed. Additional enhancement in the selectivity of laser techniques will depend on sample preparation (e.g., matrix isolation), previous selective isolation (e.g., chromatography), and/or more complex multiphoton laser techniques.

Prominent future areas for application of laser techniques to analytical fluorescence surely include transient phenomena, where the unique capability of the laser for temporal and spatial resolution is most evident. Identification of unknown transient species will be difficult, however, and will require extensive data acquisition (e.g., array detectors to generate time-resolved excitation–emission matrices[80,81]). Enhanced temporal resolution will also suggest the use of fluorescence quenching as an analytical method.[82] Such a method would be especially powerful if coupled with steady-state and time-resolved photoacoustic data.

Finally, laser techniques will permit real time *in situ* analyses that were not possible with conventional sources. Line-of-sight pollution monitoring is one example, but the explosion in fiber optics heralds new frontiers for remote analysis by analytical molecular fluorescence spectroscopy. The ultimate analytical instrument will be quite sophisticated, employing various analytical techniques, but laser-induced molecular fluorescence will certainly be one of the prominent techniques.

REFERENCES

1. J. I. Steinfeld and M. S. Wrighton, "The Laser Revolution in Energy Related Chemistry," a National Science Foundation Workshop, Massachusetts Institute of Technology (May 1976).
2. A. B. Harvey, *Anal. Chem.* **50**, 905A–912A (1978).
3. G. M. Hieftje and T. R. Copeland, *Anal. Chem.* **50**, 300R–327R (1978).
4. P. Dhamelincourt, F. Wallart, M. Leclercq, A. T. N'Guyen, and D. O. Landon, *Anal. Chem.* **51**, 414A–421A (1979).
5. G. J. Diebold and R. N. Zare, *Science* **196**, 1439–1441 (1977).
6. J. L. Hall, *Science* **202**, 147–156 (1978).
7. M. Hanabusa, C. C. Wang, S. Japar, D. K. Killinger, and W. Fisher, *J. Chem. Phys.* **66**, 2118–2120 (1977).
8. A. W. Tucker, M. Birnbaum, and C. L. Fincher, *Appl. Opt.* **14**, 1418–1422 (1975).
9. J. H. Eberly, W. C. McColgin, K. Kawaoka, and A. P. Marchetti, *Nature (London)* **215**, 215–217 (1974).
10. J. C. Wright and F. J. Gustafson, *Anal. Chem.* **50**, 1147A–1160A (1978).
11. F. P. Schäfer, ed., *Dye Lasers, Topics in Applied Physics*, Vol. 1 (Springer-Verlag, New York. 1973).
12. H. W. Latz, in *Modern Fluorescence Spectroscopy*, Vol. 1, E. L. Wehry, ed. (Plenum Press, New York, 1976) pp. 83–120.
13. K. Kato, *Appl. Phys. Lett.* **33**, 509–510 (1979).
14. J. A. Paisner and S. Hargrove, *Laser Focus* **15** (5), 38–39 (1979).
15. L. Cramer and K. G. Spears, *J. Am. Chem. Soc.* **100**, 221–227 (1978).
16. G. M. Hieftje, G. R. Haugen, and J. M. Ramsey, *Appl. Phys. Lett.* **30**, 463–466 (1977).
17. D. V. O'Connor, W. R. Ware, and J. C. Andre, *J. Phys. Chem.* **83**, 1333–1343 (1979).
18. U. P. Wild, A. R. Holzwarth, and H. P. Good, *Rev. Sci. Instrum.* **48**, 1621–1627 (1977).
19. J. H. Richardson and M. E. Ando, *Anal. Chem.* **49**, 955–959 (1977).
20. F. E. Lytle and M. S. Kelsey, *Anal. Chem.* **46**, 855–860 (1974).
21. S. L. Shapiro, ed., *Ultrashort Light Pulses, Picosecond Techniques and Applications, Topics in Applied Physics*, Vol. 18 (Springer-Verlag, New York, 1977).
22. I. S. Ruddock and D. J. Bradley, *Appl. Phys. Lett.* **29**, 196–197 (1976).
23. C. V. Shank and E. P. Ippen, *Appl. Phys. Lett.* **24**, 373–375 (1974).
24. Z. A. Yasa, A. Dienes, and J. R. Whinnery, *Appl. Phys. Lett.* **30**, 24–26 (1977).
25. L. L. Steinmetz, J. H. Richardson, and B. W. Wallin, *Appl. Phys. Lett.* **32**, 163–165 (1978).
26. J. M. Harris, R. W. Chrisman, and F. E. Lytle, *Appl. Phys. Lett.* **26**, 16–18 (1975).
27. L. L. Steinmetz, J. H. Richardson, B. W. Wallin, and W. A. Bookless, in *Picosecond Phenomena, Springer Series in Chemical Physics*, Vol. 4, C. V. Shank, E. P. Ippen, and S. L. Shapiro, eds. (Springer-Verlag, New York, 1978), pp. 67–70.
28. J. H. Richardson, L. L. Steinmetz, S. B. Deutscher, W. A. Bookless, and W. L. Schmelzinger, *Z. Naturforsch.* **33a**, 1592–1593 (1978).
29. J. H. Richardson, L. L. Steinmetz, S. B. Deutscher, W. A. Bookless, and W. L. Schmelzinger, *Anal. Biochem.* **97**, 1723 (1979).
30. E. P. Ippen and C. V. Shank, in *Picosecond Phenomena, Springer Series in Chemical Physics*, Vol. 4, C. V. Shank, E. P. Ippen, and S. L. Shapiro, eds. (Springer-Verlag, New York, 1978), pp. 103–107.
31. J. C. Mialocq and P. Goujon, *Appl. Phys. Lett.* **33**, 819–820 (1978).
32. J. J. Ewing, in *Chemical and Biochemical Applications of Lasers*, Vol. 2, C. B. Moore, ed. (Academic Press, New York, 1977), pp. 241–278.
33. R. D. Reid, J. R. McNeil, and G. J. Collins, *Appl. Phys. Lett.* **29**, 666–668 (1976).

34. D. C. Harrington and H. V. Malmstadt, *Anal. Chem.* **47**, 271–276 (1975).
35. J. M. Fitzgerald, in *Modern Fluorescence Spectroscopy, Vol. 1*, E. L. Wehry, ed. (Plenum Press, New York, 1976), pp. 45–63.
36. G. M. Klauminzer, *Laser Focus* **11** (11), 35–39 (1975).
37. J. H. Richardson and S. M. George, *Anal. Chem.* **50**, 616–620 (1978).
38. G. R. Haugen, G. M. Hieftje, and F. E. Lytle, Analytical Utility of Time-Correlated Single-Photon Counting: Elimination of Background Luminescence by Time-Filtering, paper presented at the 32nd Annual Analytical Summer Symposium, Lasers and Analytical Chemistry, Purdue University (June 1979).
39. V. J. Koester and R. M. Dowben, *Rev. Sci. Instrum.* **49**, 1186–1191 (1978).
40. B. I. Greene, R. M. Hochstrasser, and R. B. Weisman, *J. Chem. Phys.* **70**, 1247–1259 (1979).
41. W. Yu, F. Pellegrino, M. Grant, and R. R. Alfano, *J. Chem. Phys.* **67**, 1766–1773 (1977).
42. G. R. Fleming, J. M. Morris, R. J. Robbins, G. J. Woolfe, P. J. Thistlethwaite, and G. W. Robinson, *Proc. Nat. Acad. Sci. U.S.A.* **75**, 4652–4656 (1978).
43. G. W. Robinson, R. J. Robbins, G. R. Fleming, J. M. Morris, A. E. W. Knight, and R. J. S. Morrison, *J. Am. Chem. Soc.* **100**, 7145–7150 (1978).
44. J. H. Clark, S. L. Shapiro, A. J. Campillo, and K. R. Winn, *J. Am. Chem. Soc.* **101**, 746–748 (1979).
45. International Agency for Research on Cancer, *IARC Monographs on the Evaluation of Carcinogenic Risk of Chemicals to Man, Vol. 3, Certain Polycyclic Aromatic Hydrocarbons and Heterocyclic Compounds* (IARC, Lyon, 1973).
46. J. B. Birks, *Photophysics of Aromatic Molecules* (Wiley-Interscience, New York, 1970).
47. F. P. Schwarz and S. P. Wasik, *Anal. Chem.* **48**, 524–528 (1976).
48. J. H. Richardson, S. M. George, and M. E. Ando, Trace Organic Analysis: A New Frontier in Analytical Chemistry, Proceedings of the 9th Materials Research Symposium, April 1978, held at NBS, Gaithersburg, Maryland, *National Bureau of Standards Special Publication 519* (April 1979), pp. 691–696.
49. B. S. Das and G. H. Thomas, *Anal. Chem.* **50**, 967–973 (1978).
50. J. H. Richardson, K. M. Larsen, G. R. Haugen, D. C. Johnson, and J. E. Clarkson, *Anal. Chim. Acta* **116**, 407–411 (1980).
51. M. R. Berman and R. N. Zare, *Anal. Chem.* **47**, 1200–1201 (1975).
52. G. J. Diebold and R. N. Zare, in *New Applications of Lasers to Chemistry, ACS Symposium Series, Vol. 85*, G. M. Hieftje, ed. (American Chemical Society, Washington, D.C., 1978), pp. 80–90.
53. E. L. Wehry and G. Mamantov, *Anal. Chem.* **51**, 643A–656A (1979).
54. P. Tokousbalides, E. R. Hinton, Jr., R. B. Dickinson, Jr., P. V. Bilotta, E. L. Wehry, and G. Mamantov, *Anal. Chem.* **50**, 1189–1193 (1978).
55. J. R. Maple, E. L. Wehry, and G. Mamantov, *Anal. Chem.* **52**, 920–924 (1980).
56. B. Dellinger, D. S. King, R. M. Hochstrasser, and A. B. Smith III, *J. Am. Chem. Soc.* **99**, 7138–7142 (1977).
57. J. C. Brown, M. C. Edelson, and G. J. Small, *Anal. Chem.* **50**, 1394–1397 (1978).
58. R. B. Dickinson, Jr., and E. L. Wehry, *Anal. Chem.* **51**, 778–780 (1979).
59. J. H. Richardson, B. W. Wallin, D. C. Johnson, and L. W. Hrubesh, *Anal. Chim. Acta* **86**, 263–267 (1976).
60. J. H. Richardson, *Anal. Biochem.* **83**, 754–762 (1977).
61. J. H. Richardson, in *Methods in Enzymology, Vol. 66*, D. B. McCormick and L. D. Wright, eds. (Academic, New York, 1980), pp. 416–425.
62. R. F. Chen and H. Edelhoch, eds., *Biochemical Fluorescence: Concepts* (Marcel Dekker, New York, 1975).

63. C. M. O'Donnell and S. C. Suffin, *Anal. Chem.* **51**, 33A–40A (1979).
64. A. B. Bradley and R. N. Zare, *J. Am. Chem. Soc.* **98**, 620–621 (1976).
65. T. Imasaka, H. Kadone, T. Ogawa, and N. Ishibashi, *Anal. Chem.* **49**, 667–668 (1977)
66. T. Hirschfeld, *Appl. Opt.* **15**, 2965–2966 (1976); **15**, 3135–3139 (1976).
67. S. D. Lidofsky, T. Imasaka, and R. N. Zare, *Anal. Chem.* **51**, 1602–1605 (1979).
68. J. D. Winefordner, in *New Applications of Lasers to Chemistry, ACS Symposium Series*, Vol. 85, G. M. Hieftje, ed. (American Chemical Society, Washington, D.C., 1978), pp. 50–79.
69. J. C. Wright, Analysis by Laser Induced Fluorescence, paper presented at the 32nd Annual Analytical Summer Symposium, Lasers and Analytical Chemistry, Purdue University (June 1979).
70. T. G. Matthews and F. E. Lytle, *Anal. Chem.* **51**, 583–585 (1979).
71. E. S. Yeung and K. Chen, *J. Chem. Phys.* **70**, 1312–1319 (1979).
72. D. Fröhlich and M. Sondergeld, *J. Phys. E* **10**, 761–766 (1977).
73. M. J. Wirth and F. E. Lytle, *Anal. Chem.* **49**, 2054–2057 (1977).
74. J. H. Richardson, J. P. Dering, D. C. Johnson, and L. W. Hrubesh, *Anal. Chem.* **52**, 982–983 (1980).
75. J. M. Harris and N. J. Dovichi, *Anal. Chem.* **52**, 695A–706A (1980).
76. C. G. Stevens, Lawrence Livermore Laboratory, Livermore, California, private communication.
77. M. Birnbaum, in *Modern Fluorescence Spectroscopy*, Vol. 1, E. L. Wehry, ed. (Plenum Press, New York, 1976), pp. 121–157.
78. T. Hirschfeld, G. R. Haugen, D. C. Johnson, and L. W. Hrubesh, Lawrence Livermore Laboratory, Livermore, California, unpublished results.
79. J. Davis, M. Feld, C. P. Robinson, J. Steinfeld, N. Turro, W. S. Watt, and J. T. Yardley, *Laser Photochemistry and Diagnostics: Recent Advances and Future Prospects*, Report on a NSF/DOE Seminar for Government Agency Representatives, Washington, D.C. (June 1979).
80. I. M. Warner, G. D. Christian, E. R. Davidson, and J. B. Callis, *Anal. Chem.* **49**, 564–573 (1977).
81. I. M. Warner, E. R. Davidson, and G. D. Christian, *Anal. Chem.* **49**, 2155–2159 (1977).
82. G. R. Haugen, J. H. Richardson, and J. E. Clarkson, Laser Induced Fluorescence: Extension to Non-fluorescent Materials, in *Proceedings of the New Concepts Symposium and Workshop on Detection and Identification of Explosives, Reston, Virginia (October 1978)*, sponsored by the U.S. Departments of Treasury, Energy, Justice, and Transportation (NTIS), pp. 249–251.

Chapter 2

A Linear Response Theory Approach to Time-Resolved Fluorometry

G. M. Hieftje and E. E. Vogelstein

A. INTRODUCTION

The phenomenon of fluorescence has been recognized since the mid-19th century, when it was first correctly described by Sir George Stokes.[1] Even the measurement of fluorescence lifetimes was successfully approached as early as 1962 by E. Gaviola.[2] However, only quite recently has instrumentation become available to permit the development of fluorometry into a useful chemical tool. In particular, the development of practical techniques for time-resolved fluorescence measurements only became possible after about 1950, when radiation detectors and signal-processing electronics with high sensitivities and nanosecond time resolution became commercially available. Since then, there has been a tremendous amount of activity in this field. Not only have several methods for the measurements of fluorescence lifetimes been well developed, but they have already been applied to many interesting problems in several areas of chemistry.

In this chapter, the measurement of luminescence lifetimes is treated from an unusual vantage point, specifically that of linear response theory. It will be shown how linear response theory can be employed to relate

G. M. Hieftje and E. E. Vogelstein • Department of Chemistry, Indiana University, Bloomington, Indiana 47405

essentially all commonly used methods for time-resolved fluorometry. Perhaps more importantly, it will be demonstrated that new approaches to subnanosecond lifetime measurement can be derived from information theory concepts. No attempt will be made to be exhaustive in the coverage of time-resolved measurement techniques, nor will an exhaustive review of the literature be provided. Rather, the goal of the chapter will be to provide a framework within which essentially all lifetime measurement techniques can be unified. To begin, let us briefly examine the nature of time-resolved fluorometry and learn how it can be viewed from the standpoint of linear response theory.

B. PRINCIPLES OF TIME-RESOLVED FLUOROMETRY

Fluorescence is a linear process (i.e., it is proportional to the power of exciting radiation), and a unified understanding of the techniques for the measurement of fluorescence lifetimes can be obtained by viewing time-resolved fluorometry from the standpoint of linear response theory. Therefore let us digress a bit to introduce the basic concepts of linear response theory.

1. Linear Response Theory

In linear response theory any system is described by specifying its *transfer function*. The transfer function completely describes the system's response to any time-dependent perturbation. These basic ideas are illustrated schematically in Figure 1. Clearly, the transfer function is itself an important system property. Moreover, various other system parameters can be determined from it.

There are two ways of specifying the transfer function of a system or network: by its *frequency response function* or by its *impulse response*

FIGURE 1. Schematic illustration of the concept of a system transfer function. $P(t)$ is the time-dependent input perturbation, T is the transfer function of the system (T is a function of system properties $A, B, \ldots$), and $R(t)$ is the response of the system as a function of time (R is a function of the perturbation P and the transfer function T).

function. The frequency response gives the attenuation and phase shift caused by the system as a function of frequency. It can be measured by applying a sinusoidal perturbation to the system's input and measuring at the system's output the relative amplitude and phase of the resulting sinusoidal response. A plot of this amplitude and phase behavior as a function of frequency then constitutes the frequency response function. A schematic illustration of the determination of a frequency response function is shown in Figure 2. Amplitude and phase fluorometry are techniques that rely upon the frequency response function, as discussed later.

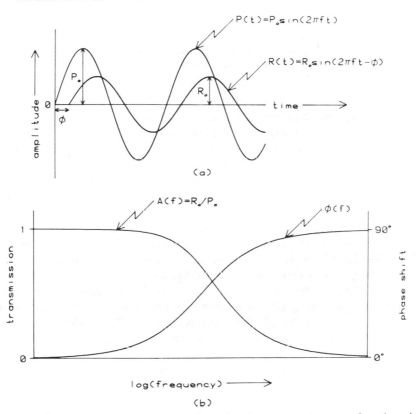

FIGURE 2. Schematic of the determination of a frequency response function. (a) Schematic illustration of the measurement of a system's response at frequency f. P_0 is the amplitude of the sinusoidal input, R_0 is the amplitude of the resulting sinusoidal output, and ϕ is the phase lag of the response $R(t)$ relative to the input perturbation $P(t)$. (b) The system's frequency response function. $A(f) = R_0/P_0$ is the amplitude of the system's response relative to the input perturbation as a function of frequency and is the amplitude portion of the frequency response function, $\phi(f)$ is the phase shift of the system's response relative to the input perturbation as a function of frequency and is the phase portion of the frequency response function.

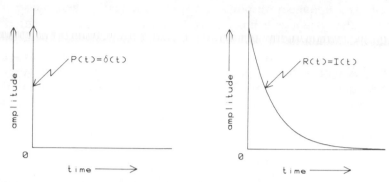

FIGURE 3. Schematic illustration of the measurement of a system's impulse response function. $P(t)$ and $R(t)$ are the particular time-dependent input and output functions to the system being perturbed. $\delta(t)$ is the delta-function input and $I(t)$ is the system's impulse response function.

The impulse response function, on the other hand, gives the time response of the system to a delta-function (impulse) perturbation. An ideal delta function can be thought of as an infinitely narrow spike of infinite amplitude. Consequently, an approximation to the impulse response function can be measured by applying a short-duration pulse of large amplitude to a system's input and monitoring the resulting time response at the system's output. Figure 3 schematically illustrates such a measurement. Notice that the impulse response is a function of *time* , unlike the frequency response. As we shall see, these two functions are mathematically related. Understandably, the impulse response function is related to the pulse techniques for measuring luminescence lifetimes.

According to Fourier theory, an ideal impulse contains all frequencies. Therefore, applying a delta-function input to a system is equivalent to simultaneously sending all frequencies into it. The impulse response function must then also contain all frequencies superimposed on each other, but each frequency component will have been attentuated and phase shifted according to the system's frequency response. Therefore, the impulse response function contains exactly the same information as the frequency response function. In fact, the two functions are entirely equivalent and can be simply related by Fourier transformation. Accordingly, if a system's impulse response function is known, its frequency components can be unscrambled by Fourier transformation to yield the frequency response function. Similarly, the impulse response function can be calculated from the frequency response function by inverse Fourier transformation.

Either the frequency or impulse response function completely determines the response of a system or device to any arbitrary input. In terms of the frequency response function, the frequency spectrum of a system's output is related to that of the corresponding input signal by

$$Y(f) = X(f)A(f) \tag{1a}$$

$$\beta(f) = \alpha(f) + \phi(f) \tag{1b}$$

In these equations $Y(f)$ and $\beta(f)$ are the amplitude and phase portions of the output frequency spectrum, $X(f)$ and $\alpha(f)$ are similarly the components of the input signal's frequency spectrum, and $\Lambda(f)$ and $\phi(f)$ are the amplitude and phase of the system's frequency response function. In words, Equation (1a) says that the amplitude of any frequency component in the output signal is equal to its amplitude in the input signal multiplied by the system's transmission at that frequency. Similarly, the phase of each frequency component in the output is just equal to its phase in the input plus the characteristic phase shift of the system at that frequency. These relations make sense intuitively; moreover, they are quite familiar for electronic systems, such as RC filter circuits and amplifiers.

In terms of the impulse response function, on the other hand, the response of a system, $R(t)$, to an input perturbation, $P(t)$, is given by the convolution of the input perturbation with the system's impulse response function $I(t)$. This convolution process is described mathematically by the relation

$$R(t) = P(t)*I(t) = \int_0^t P(T)I(T - \tau) \, dT \tag{2}$$

where the asterisk denotes convolution, t is time, and τ is just a dummy integration variable. A simple explanation of the form of the convolution integral is possible if one analyzes more closely the convolution process itself.

Convolution

Let us intuitively derive the mathematical expression for convolution by considering a series of experiments with some hypothetical test system; Fig. 4 illustrates this approach schematically. In each experiment the system is perturbed with a specific input signal, $P(t)$, and we try to describe mathematically what the system's response, $R(t)$, will be in each case.

(1) The system is perturbed at time $t = 0$ with a delta-function input of amplitude $P(0)$. The response of the system after time $t = 0$ will be, by definiton, the system's impulse response function, $I(t)$, scaled in amplitude

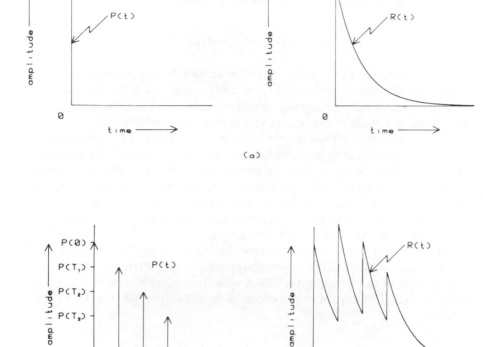

FIGURE 4. Schematic illustration of the convolution process. $P(t)$ is the amplitude–time profile of the input signal, $R(t)$ is the response of the system as a function of time and $I(t)$ is the impulse response function of the system. See text for explanation.

by that of the perturbing impulse. This experiment is illustrated schematically in Figure 4(a), and is described mathematically by

$$P_1(t) = P(0)\delta(t) \tag{3a}$$

$$R_1(t) = P(0)I(t) \tag{3b}$$

(2) The system is perturbed with a delta-function input of amplitude $P_1(t)$ at time $t = T$. The response of the system will again be just the impulse

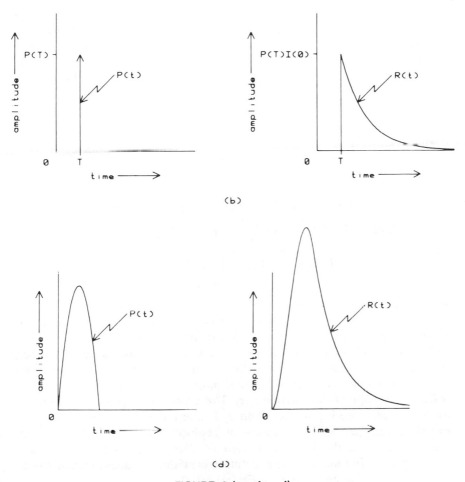

FIGURE 4 (continued)

response scaled by the amplitude of the input pulse; however, the response is now displaced in time to $t = T$. Figure 4(b) schematically illustrates this experiment, and its mathematical description is

$$P_2(t) = P_1(T)\delta(t - T) \tag{4a}$$

$$R_2(t) = P_1(T)I(t - T) \tag{4b}$$

(3) The system is perturbed with a train of delta-function pulses. Specifically, let the system be perturbed by the following series of impulses:

a delta function of amplitude $P(0)$ at time $t = 0$, a delta function of amplitude $P(T_1)$ at time $t = T_1$, a delta function of amplitude $P(T_2)$ at time $t = T_2$, etc. From the preceding discussion, each of these input impulses induces a response from the system in the form of the system's impulse response function scaled by the amplitude of the input pulse and displaced in time by the pulse's time of occurrence. The total response of the system to the train of perturbing impulses is then just the sum of the contributions from all of the impulses; i.e., the amplitude of the system's output at any time t will be equal to the sum of the amplitudes at that time of the responses induced by all preceding input impulses. This experiment is illustrated schematically in Figure 4(c) and it is described mathematically by

$$P_3(t) = P(0)\delta(t) + P(T_1)\delta(t - T_1) + P(T_2)\delta(t - T_2) + \cdots \tag{5a}$$

$$R_3(t) = P(0)I(t) + P(T_1)I(t - T_1) + P(T_2)I(t - T_2) + \cdots$$

$$+ P(T_n)I(t - T_n) = \sum_{T=0}^{t} P(T)I(t - T) \tag{5b}$$

where $T_n \leq t$ but $T_{n+1} > t$.

(4) Finally, let us consider a perturbation of the system with some real signal of finite duration, whose amplitude–time profile is given by $P(t)$ [cf. Figure 4(d)]. This input signal can be viewed as a train of delta-function pulses, just like that of the preceding example, but the interval between successive pulses has become infinitesimal and the amplitude of the pulses follows the shape of the function $P(t)$. Therefore the response of the system in this general case is also described by Equation (5b), with just one minor modification: The summation must be replaced by an integral, since successive pulses in the imaginary train of delta-function pulses are now infinitely close in time. If this substitution is made, one does in fact arrive at the convolution integral given in Equation (2).

2. Practical Considerations

From the foregoing discussion, it should be apparent that either the impulse response or frequency response function can be employed to describe the temporal response characteristics of a system we wish to test, and that both methods provide the same information. However, depending on the particular experimental situation, one method or the other might be superior. Therefore, it is appropriate to briefly consider the experimental requirements for measuring each kind of function.

Obviously, to determine an impulse response function the system under test must be perturbed with a source of energy that appears temporally as a spike. Ideally this spike or impulse should approach a mathematical delta function; that is, it should possess infinite amplitude and zero width and have an integral of unity. In practice, however, the perturbing source must emit a spike that is narrow with respect to the temporal characteristics of the transfer function being measured. Moreover, the maximum amplitude of the spike will be limited practically by the linearity of the system being tested, while the minimum amplitude will be bounded by the sensitivity of the system to the perturbation and the required magnitude of the response it elicits. Measurement of the elicited impulse response function simply requires a detector capable of monitoring the response and of faithfully following its temporal behavior.

Obviously, severe constraints exist in the choice of a real perturbation source for impulse response determinations. Often, the perturbing source is not brief enough in duration, which results in a distorted response trace. Alternatively, the source might be too weak to evoke a measurable response, or so intense that nonlinear behavior is produced. As will be indicated later, all these limitations exist in time-resolved fluorometry and great care must be taken to overcome or avoid them.

In the measurement of a frequency response function, several alternative approaches exist. The simplest such approach, and one that produces the highest signal-to-noise ratio, simply involves perturbing the system under test with a sinusoidally modulated energy source and observing the resulting response with a tuned, phase-sensitive detector. A linked frequency sweep of source and detector then provides both parts (phase and amplitude) of the frequency response function. Alternatively, several fixed-frequency sources can perturb the system simultaneously, with the resulting response being examined either by a series of individual frequency and phase-sensitive detectors or with a single frequency-swept device. This latter kind of device is termed a "spectrum analyzer." Of course, the second method for frequency response measurement results in measurements at only specific, discrete frequencies and limits the resolution of the resulting response curve.

To improve resolution on the curve, additional perturbing sources could be added, with an attendant increase in complexity. In the limit one could envision an infinite number of discrete sources, enabling the entire spectrum to be resolved. Interestingly, this seemingly exotic limit is relatively simple to implement practically. White-noise sources are available that emit energy over enormous frequency ranges (dc to GHz) and provide outputs that, for most purposes, appear to have originated from just such an infinite array of sources. Thus, with a white-noise source, a system can be perturbed simultaneously over a broad frequency range, with the resulting response being simply monitored by a swept-frequency spectrum analyzer.

Frequency response measurements, be they swept, multifrequency, or noise input approaches, are often superior to impulse measurements. Whereas the impulse requires all energy that elicits a measured response to be concentrated in a single instant, the perturbing energy can be spread out in time when the frequency domain is employed. This capability often results in greater detectability, signal-to-noise ratios, and less danger from saturation (nonlinearity) than is found in the pulse techniques. As a result of this limitation engineers employing linear response theory have devised a simple way of obtaining impulse response functions, using multifrequency or white-noise perturbing sources. The basic idea behind these measurements is outlined in Figure 5.

As shown in Figure 5 (top), there is a fixed mathematical relationship between white noise and a delta function. Specifically, the delta function is obtained by *autocorrelating* the noise. That this result arises can be appreciated from the basic nature of the correlation process.[3] In particular, the process of autocorrelation produces a phase registration of all of the frequency components in any complex (nonsinusoidal) waveform. That is, each frequency component in the waveform is converted to a cosine wave and therefore has its maximum value at a location equal to zero on the horizontal axis. Consequently, *all* frequency components in the complex waveform align at this phase-related point, causing their amplitudes to add. At all other points on the horizontal axis, however, the different frequency components in the original waveform will be out of phase by varying degrees and, when summed, will produce a resultant that is probably lower than the amplitude of the "zero" point. For a waveform similar to white noise, which

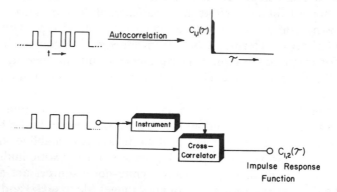

FIGURE 5. Generation of an impulse response function using a noisy perturbing source. (Top) Autocorrelation of a random waveform produces a delta function. $C_{1,1}(\tau)$ is the autocorrelation function and τ is the autocorrelation displacement. (Bottom) Cross-correlation of a random perturbation with the response it elicits to generate the impulse response function. $C_{1,2}(\tau)$ is the cross-correlation function.

contains *all frequencies*, there will be an infinite number of such waves, so that for every one that has a large positive value at some horizontal axis location, there will be another that has an equal but negative amplitude, causing the resultant to be zero. Consequently, for a completely random waveform such as white noise, all frequency components add only at the "zero" point and produce everywhere else a value of zero.

How these concepts can be applied to the measurement of an impulse response function is illustrated in Figure 5 (bottom). In this case white noise is not correlated with itself (autocorrelation), but rather is correlated with a version of itself that has been passed through the device we wish to test. Intuitively, one would imagine that the random waveform would be altered upon passing through this device; in particular, that some of its frequency components would have been attenuated or phase shifted somewhat, in accordance with the system's frequency response function. For simplicity, let us assume here that this system attenuates the high-frequency components in the random waveform. As a consequence, one would expect these high-frequency components also to be absent from the cross-correlation function calculated between the two waves. Loss of these high-frequency components would strip the impulse response function of its rapidly changing features and would yield a smoothed waveform rather than the delta function that would otherwise have been produced. Although this discussion has centered on the *transmission* of a random waveform through a device to be tested, it should be clear that the same considerations apply to a response elicited by such a waveform.

This alternative scheme for the measurement of impulse response functions has several attractive features. Like the frequency response measurements, it applies the perturbation to the system under test over an extended time period, rather than all at once. The system is thus less likely to be strained or driven into nonlinear response. In addition, this alternative method is relatively immune to a constant offset in the input perturbation. This feature is a substantial advantage in some applications. For example, the impulse response function of a large electrically driven turbine can be obtained *while the turbine is running* simply by adding a small random electrical variation to the constant drive current and by cross-correlating that random waveform with the minute fluctuations in the turbine's velocity that result. Finally, this method can prove advantageous just because of the ubiquitous nature of white-noise or random processes. Recognizing that most things we examine are perturbed by the natural, often stochastic events in their environment, scientists have devised extremely clever ways to measure response functions without the need for external perturbation.[4-7]

Many examples could be cited where impulse or frequency response functions are used in science. Because of the simplicity of their interpretation, the impulse techniques are the most widely employed. For example, Fourier-transform nuclear magnetic resonance spectrometry is essentially an impulse technique, wherein a pulse of radio-frequency energy is used to elicit a response from nuclei (their free induction decay). Impulse methods are also used in perturbation techniques for the measurement of rapid kinetics (e.g., pressure-jump and temperature-jump approaches). Finally, analytical gas and liquid chromatography are impulse-type measurements and involve the application of a pulse of sample material, which yields as an impulse response function the desired chromatogram.

These examples hopefully reveal the importance and scope of impulse and frequency response measurements and also suggest that perturbing sources need not be electrical in nature but might consist of energy input in the form of a light pulse, a temperature variation, or an increment of chemical sample. Let us now turn our attention to an area in which these approaches become especially useful—time-resolved fluorometry.

C. LINEAR RESPONSE FUNCTIONS IN TIME-RESOLVED FLUOROMETRY

How can the concepts of linear response theory be applied to time-resolved fluorometry? In such an application, the generalized system discussed in the preceding section is specifically the fluorescent sample under study. Applying a time-dependent perturbation correspondingly translates into illuminating the sample with a source whose output radiance varies as a function of time. Similarly, monitoring the response of the system to the perturbation now implies measuring the time-dependent intensity of the resulting fluorescence. Again, the response and the input perturbation are related via the system's impulse response (transfer) function. For a fluorescent sample, this transfer function is determined by the decay kinetics of the probed excited state.

The usual objectives of time-resolved fluorescence studies are the measurement of excited-state lifetimes and the study of the decay kinetics of species in excited states. From the standpoint of linear response theory, an experiment in time-resolved fluorometry involves measuring the transfer (impulse response) function of the molecular system being studied; from this function the excited-state lifetime or decay-rate constant of interest can be calculated.

Because there are two ways of specifying a system's transfer function, there are two classical approaches to time-resolved fluorometry. The pulse techniques for the measurement of fluorescence lifetimes[8-10] constitute one approach. With these techniques it is the impulse response function of the molecular system being studied that is measured. The other classical approach to time-resolved fluorometry employs the measurement of points on the frequency response function of the fluorescent system. This approach includes the modulation and phase-shift[8,11,12] techniques for the measurement of fluorescence lifetimes. Finally, a group of new techniques has recently arisen that revolves around the use of a stochastic or noisy perturbation and either cross-correlation or spectrum analysis to generate, respectively, the impulse response function or frequency response function of a fluorophore.

Let us now take a closer look at these approaches to time-resolved fluorometry. In particular, let us consider what the impulse and frequency response functions of a fluorescent sample look like and how they are related to the excited-state lifetime of the fluorophore. Also to be discussed are the experimental measurements involved and how lifetime values can be calculated from the data in each case. Finally, new methods based on correlation will be examined and their capabilities compared with the classical schemes.

1. Pulse Fluorometry

In the pulse techniques for the measurement of fluorescence lifetimes, the sample of interest is illuminated with an intense, brief pulse of light and the intensity of the resulting fluorescence is recorded as a function of time. If the exciting light pulse is sufficiently short, the measured fluorescence decay curve will be a good approximation to the sample's impulse response function.

In general, several processes contribute to the relaxation of molecules in solution from an excited singlet state: the radiative process of fluorescence, as well as such nonradiative processes as internal conversion, intersystem crossing, and quenching by other species present in the solution. Fluorescence, internal conversion, and intersystem crossing are unimolecular processes; therefore they follow simple first-order kinetics. In addition, quenching by other species, although a bimolecular process, can in most cases be described in terms of pseudo-first-order kinetics. Therefore, for most fluorescent samples, the excited-state population established by an impulse of exciting light will decay exponentially according to the familiar decay law for first-order kinetics. The rate of this decay is determined by the sum of the rates of all contributing relaxation processes. Because the intensity of fluorescence from a sample reflects its excited-state population, it follows that the impulse response function of such a sample is just a simple decaying exponential; moreover, the decay constant of this function directly

gives the overall relaxation rate for the probed excited state of the fluorophore.

To be more quantitative, for most fluorescent samples the impulse
response function, $I(t)$, will have the form

$$I(t) = I_0 \, e^{-kt} \tag{6}$$

where I_0 is just a scaling factor representing the peak fluorescence intensity
and k represents the overall relaxation rate for the probed excited state. The
value of k, in turn, is given by

$$k = k_F + k_{IC} + k_{IX} + k_Q \tag{7}$$

where k_F, k_{IC}, and k_{IX} are the first-order rate constants for fluorescence,
internal conversion, and intersystem crossing from the probed state,
respectively, and k_Q is the pseudo-first-order rate constant for bimolecular
quenching of that state.

Often the decay kinetics of an excited state are described in terms of its
lifetime. The fluorescence lifetime of an excited state is by definition the time
required for the excited-state population is decay to $1/e$, or $\sim37\%$, of its
initial value, following excitation by an impulse of light. Therefore, the
lifetime, τ, is simply the time corresponding to the $1/e$ point of the
fluorophore's impulse response function. From Equation (6) it follows that

$$\tau = 1/k \tag{8}$$

which indicates that the lifetime of an excited state gives the reciprocal of the
state's overall decay-rate constant.

Clearly, for any fluorescent sample that exhibits such exponential decay
behavior, the lifetime of the probed excited state can easily be calculated
from the sample's impulse response function, i.e., its experimental fluorescence decay curve. In practice, one can simply prepare a semilogarithmic
plot of the fluorescence decay data and determine the slope of the resulting
straight line. Taking the logarithm of Equation (6), one obtains

$$\log I(t) = -kt + \log I_0 \tag{9}$$

which represents a straight line with slope k. This procedure constitutes the
so-called *graphical slope method* for the evaluation of fluorescence lifetimes
from pulse fluorometric data.

Although most fluorescent samples exhibit the simple exponential
decay behavior discussed above, there are some important exceptions.

Specifically, any sample that contains more than one fluorescent species will exhibit nonexponential decay in pulse fluorometry. For such a sample, the impulse response function is a sum of exponential terms, with one term for each fluorescent species present. In general, it then becomes difficult to extract valid excited-state lifetime values for the individual components from the sample's fluorescence decay function.

In addition, not always can bimolecular quenching be described in terms of pseudo-first-order kinetics. Therefore, some one-component fluorescent samples might also exhibit nonexponential decay behavior. Study of the form of the fluorescence decay function for such samples can, however, reveal much about the mechanism of the quenching process involved.

The relaxation rates for excited singlet states are often greater than $10^9 \sec^{-1}$; accordingly, fluorescence lifetimes fall in the nanosecond and subnanosecond time range. Therefore, the measurement of fluorescence lifetimes by pulse techniques poses some difficult instrumental problems: sources capable of generating nanosecond or subnanosecond light pulses and sensitive detectors and signal-processing electronics capable of responding on this time scale are needed. Significantly, many fluorophores have excited-state lifetimes of the same order or even shorter than the time-resolution capabilities of available instrumentation. If the duration of the light pulses provided by the excitation source is not negligible relative to the lifetime of the fluorophore being studied, the sample's measured fluorescence will not be an accurate representation of its impulse response function. Similarly, if the response of the detection system is limited in the time range of interest, then the measured fluorescence signal will be distorted. In such cases, the measured fluorescence decay curve will be the convolution of the excitation light pulse and the impulse response of the detection system with the desired impulse response function of the sample. Therefore, the determination of very short fluorescence lifetimes requires *deconvolution* analysis[13] in order to extract the true impulse response function of the sample from the experimentally measured fluorescence decay curve.

2. Modulation and Phase Fluorometry

Let us now consider what the frequency response function of a fluorescent sample looks like. Recall from the earlier discussion that a system's impulse and frequency response functions can be simply related by Fourier transformation. Therefore, for any fluorescent sample that exhibits simple exponential decay behavior in pulse fluorometry, the frequency response function is just a Lorentzian, i.e., the Fourier transform of an exponential. The amplitude, $A(f)$, and phase, $\phi(f)$, of a Lorentzian obtained by Fourier transformation of an exponential with decay time, τ, are given by

$$A(f) = \frac{1}{(1 + 4\pi^2 f^2 \tau^2)^{1/2}} \tag{10a}$$

$$\phi(f) = \arctan(2\pi f\tau) \tag{10b}$$

The relationship of a fluorophore's frequency response function to its excited-state lifetime τ can be expressed in terms of the frequency at the half-maximum point of either the amplitude or phase portion of the frequency response function. From Equations (10a) and (10b) one obtains the following relations:

$$\tau = \frac{3^{1/2}}{2\pi} \frac{1}{f_{1/2}} \tag{11a}$$

$$\tau = \frac{1}{2\pi} \frac{1}{f_{45°}} \tag{11b}$$

where $f_{1/2}$ denotes the frequency at the half-maximum point of the amplitude response function [i.e., the frequency at which $A(f) = 1/2$], and $f_{45°}$ is the frequency at the half-maximum point of the phase portion of the frequency response function [i.e., the frequency at which $\phi(f) = 45°$]. Notice that τ is simply the reciprocal of either $f_{1/2}$ or $f_{45°}$ times a constant. A Lorentzian frequency response function and its relationship to the excited-state lifetime of the corresponding fluorophore are illustrated in Figure 6.

In the conventional modulation and phase-shift techniques for the measurement of fluorescence lifetimes, the sample of interest is illuminated with a source whose output is sinusoidally modulated at some frequency. The resulting fluorescence is observed and will also be sinusoidally modulated at the source frequency. In the modulation techniques, the modulation depth of the fluorescence relative to the exciting light is determined. In the phase-shift techniques, on the other hand, the phase shift of the modulated fluorescence relative to the exciting light is measured. From the standpoint of linear response theory these techniques simply involve the determination of single points on the frequency response function of the sample. In either case the excited-state lifetime of the fluorophore can easily be calculated from the experimental data.

Modulation fluorometry involves the measurement of one point on the sample's amplitude response function, namely, the value of $A(f)$ at the

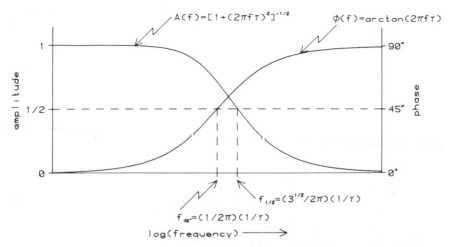

FIGURE 6. Illustration of the amplitude $A(f)$ and phase $\phi(f)$ of the Lorentzian frequency response function of a fluorophore with excited-state lifetime τ.

particular modulation frequency used in the experiment. Therefore, solution of Equation (10a) yields the expression

$$\tau = \frac{1}{2\pi f_0} \left(\frac{1}{A_0^2} - 1 \right)^{1/2} \tag{12a}$$

for the excited-state lifetime of the fluorophore, where A_0 is the measured relative modulation depth of the fluorescence and f_0 is the modulation frequency. In phase fluorometry, on the other hand, the value of $\phi(f)$ is measured at the modulation frequency of the experiment, i.e., one point on the phase portion of the sample's frequency response function is determined. In this case, τ can be calculated from the measured phase shift, ϕ_0, between the modulated fluorescence and the exciting light from Equation (10b), viz.,

$$\tau = \frac{1}{2\pi f_0} \tan \phi_0 \tag{12b}$$

where f_0 again represents the particular modulation frequency used in the experiment.

Note that the Lorentzian frequency response illustrated in Figure 6 agrees well with our intuitive feeling for how a fluorophore should respond to modulated excited light. At low modulation frequencies we would expect the excited-state population in the sample, and therefore the intensity of sample fluorescence, to closely follow the variations in exciting light. Thus, there should be little attenuation or phase shift of the modulated fluorescence compared to the exciting light. This behavior is reflected in the Lorentzian frequency response function by amplitude values near unity and phase values near zero at low frequencies. In contrast, at higher modulation frequencies we would expect the excited-state population in the sample to no longer faithfully follow the variations in exciting light because of the finite lifetime of the excited state. This behavior is seen as a phase lag and decreased modulation depth in the observed fluorescence. Accordingly, the phase of the Lorentzian frequency response function increases from 0° toward 90°, and its amplitude decreases from 1 toward 0 as frequency increases. Furthermore, we would expect that a fluorophore with a short excited-state lifetime could more closely follow higher modulation frequencies than could a fluorophore with a longer excited-state lifetime. In fact (as stated earlier), the frequencies at the half-maximum points of the amplitude and phase spectra (cf. Figure 6) are inversely proportional to the excited-state lifetime of the fluorophore.

Because fluorescence lifetimes are ordinarily very short, high modulation frequencies are required to provide a measurable phase shift or attenuation in the detected fluorescence signal. Accordingly, the measurement of fluorescence lifetimes by modulation and phase-shift techniques can best be accomplished with light modulated in the megahertz to gigahertz frequency range. Therefore, this approach to the measurement of fluorescence lifetimes, like the pulse method, requires sophisticated instrumentation.

As indicated earlier, a phase-shift or amplitude measurement at a particular modulation frequency corresponds to the determination of just one point on the frequency response function of the fluorescent sample. In theory, the whole response function of the sample could be determined by repeating these measurements over a broad range of modulation frequencies. In practice, however, the difficulties associated with light modulation at frequencies in the megahertz–gigahertz range have limited conventional modulation and phase fluorometers to the use of only one, two, or three discrete modulation frequencies. If the sample of interest is known to exhibit simple exponential decay behavior, this limitation is not serious. However, nonexponential decay behavior, i.e., deviation of the frequency response function from the Lorentzian form, is not detectable from a single phase or amplitude measurement. If measurements are made at two or three modulation frequencies, deviation from Lorentzian frequency response character is still defined by only two or three data points. Therefore, study of the decay kinetics of samples that exhibit nonexponential decay behavior is difficult with conventional phase or modulation fluorometers. Recently, a modulation fluorometer has been constructed based on a cw laser as a multi-frequency-modulated source.[14,15] This instrument should be applicable to the study of complex fluorescence decay behavior, for it enables amplitude measurements to be made at a large number of frequencies, thus accurately defining the sample's entire frequency response function.

Although pulse fluorometry is better suited for the study of complex decay behavior than are modulation or phase fluorometry, these latter techniques possess advantages in the measurement of very short fluorescence lifetimes. Because the determination of such short lifetimes requires the use of extremely high modulation frequencies, the modulation of the measured fluorescence signal is usually distorted by the limited frequency response of the detection system being used. Conveniently, correction for the response characteristics of the detection system is straightforward in modulation and phase fluorometry. In the frequency domain (unlike in the time domain), "deconvolution" involves a simple division [in modulation fluorometry cf. Equation (1a)] or subtraction [in phase fluorometry cf. Equation (1b)]. In contrast, pulse fluorometry requires correction for instrument characteristics by actual deconvolution analysis [cf. Equation (2)].

3. Correlation-Based Methods in Time-Resolved Fluorometry

As indicated earlier, either the impulse or frequency response functions can be generated through use of a randomly varying perturbation source. If the impulse response function is desired, the source's fluctuations must be cross-correlated with the elicited response; if the frequency response is desired, it is only necessary to employ a swept-frequency analyzer. Let us now examine how these concepts can be applied to the measurement of luminescence lifetimes.

Figure 7 illustrates schematically the kind of instrumentation that would be required to obtain a luminescence lifetime using the correlation approach. As before, the system being perturbed is the fluorescent species under study; its perturbation is a time-varying light flux and its observed response to this perturbation is the observed luminescence. However, in the present case the light flux is randomly modulated either internally or by means of an external modulator as shown. Excitation and emission monochromators isolate the wavelengths of interest and a high-speed photodetector converts the fluctuating luminescence to a proportional current for processing. This processing involves either the measurement of a frequency response function by means of a spectrum analyzer or an impulse response (decay curve) with a cross-correlation computer.

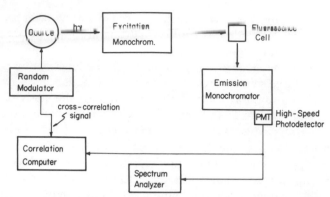

FIGURE 7. Hypothetical instrument for measuring luminescence lifetimes using a randomly modulated light source.

Unfortunately, the hypothetical instrument in Figure 7 cannot be simply implemented in the laboratory. For the displayed impulse or frequency response function to accurately reflect the kinetics (time dependence) of the fluorophore under investigation, the correlation computer must be capable of displaying subnanosecond-level data, the spectrum analyzer must be able to register frequencies in the gigahertz regime, and the light source must be randomly modulated over a frequency range from dc to several gigahertz. If these conditions are not met, extensive deconvolution procedures will be necessary and the resulting calculated decay kinetics will be less precise. Let us then examine the kinds of devices that might be realistically employed for, respectively, the source, spectrum analyzer, and correlation computer.

For the source, what is needed is a device whose emitted flux varies randomly over a broad frequency range. Obviously, the simplest such device is a cw lamp. Although we ordinarily consider such a lamp to be just a source of dc light, its beam is actually comprised of a photon flux; the random nature of photon emission and detection produces in the source emission a white-noise character. Unfortunately, this noisy component is a relatively small fraction of the dc level of the source and the dc level contributes strongly to the shot noise produced upon detection of the beam. Consequently, calculations show that a luminescence decay curve obtained through use of this source in a cross-correlation scheme would exhibit at best a signal-to-noise ratio of unity, hardly a desirable situation.

To generate higher signal-to-noise ratios, what is needed is a source whose amplitude is modulated at a depth greater than that of a cw lamp. Included in such sources might be free-running or randomly driven flash lamps, electro-optically or acousto-optically modulated cw light sources, and cw lasers. This latter source deserves further comment and is the one most often employed in this cross-correlation approach.

Noise in a cw laser arises not only from the random generation and arrival of photons, but from a phenomenon termed "mode noise." Mode noise originates in a laser's spectral structure and results from the beating of individual modes with each other. It will be recalled that modes in a laser are separated by $c/2L$, where c is the speed of light and L is the laser's cavity length (the distance between the laser's mirrors). These modes will extend over the entire emission bandwidth of the lasing medium; for an argon-ion laser this range is approximately 4 GHz, whereas for a dye laser it can be larger than 200 GHz. The beating of these modes with each other then produces a series of frequencies, spaced by $c/2L$, at which the laser is simultaneously modulated. Moreover, these discrete frequencies will extend as high as the emission bandwidth of the lasing medium (i.e., 4 GHz for an argon-ion laser). Thus, for a 1-m argon-ion laser, $c/2L$ will be 150 MHz, and the laser will appear to be amplitude modulated at all discrete frequencies from 150 MHz to approximately 4 GHz, in increments of 150 MHz. The laser will therefore not be a truly random source, but its output will appear to be random and will in fact be modulated over a sufficiently broad frequency range.

The use of laser mode noise in a frequency response function approach to determining luminescence lifetimes was recently described[14,15]; a schematic diagram of the instrument is shown in Figure 8. In this work, a cw laser served as a multifrequency modulated source and a spectrum analyzer was employed to display the frequency response function of several fluorophores that were investigated. From Figure 8 one can derive an intuitive feeling for how this scheme works. The top spectrum on the right side of Figure 8 displays the spectrum analyzer output expected if the sampled cell contained simply a scattering medium. In such a case, the spectrum would indicate the amplitude of fluctuations in the laser's output. As described above, these fluctuations extend to extremely high frequencies and are present at discrete intervals. In contrast, the bottom spectrum in Figure 8 indicates the spectrum analyzer output when the scattering medium is replaced by a fluorophore capable of being excited by the fluctuating laser radiation. To understand this spectrum, recall that the laser is, in essence, self-amplitude-modulated at each of the discrete frequencies indicated in the upper spectrum. Obviously, the excited state of the fluorophore can follow the slowest of these fluctuations quite faithfully and will accordingly yield a strong amplitude modulation in fluorescence at those frequencies. However, the finite excited-state lifetime of the fluorophore prevents it from fluctuating at the highest laser beat frequencies, to produce a gradual roll-off. As described in the section on linear response theory, this roll-off

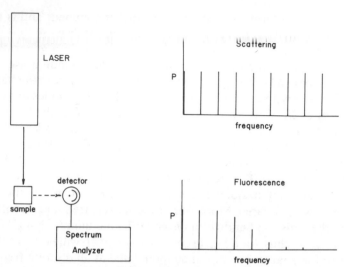

FIGURE 8. Obtaining luminescence lifetimes using a cw laser as a multifrequency modulated light source. Top plot illustrates laser mode beats, which occur at discrete frequencies $= c/2L$ (c is the speed of light; L is the laser cavity length). Bottom plot illustrates attenuation of mode beats at high frequencies, which is caused by finite lifetime of fluorophore. See text for discussion.

should follow a Lorentzian pattern, betraying the exponential excited-state decay behavior.

In reality, displayed spectra are never as clean as those shown schematically in Figure 8. Real plots, reproduced in Figure 9, reveal that the discrete fluctuations in laser radiation are not all equal in amplitude, requiring the beat-frequency spectra from fluorophores to be normalized by the amplitudes of individual beat frequencies. Thus, the two right-hand spectra in Figure 9 would be divided by that shown in Figure 9A. As suggested earlier, this division is especially meaningful here, since data are displayed in the frequency domain. Therefore, this normalization actually constitutes deconvolution.

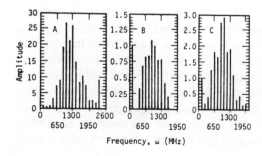

FIGURE 9. Mode-beat plots obtained with the instrument shown in Figure 8. (A) Mode beats in laser, measured with a scattering suspension in the sample cuvette. (B) Fluorescence beat plot obtained from Rhodamine B. (C). Plot corresponding to Rose Bengal. From Hieftje et al.,[14] reproduced by permission of the American Institute of Physics.

Normalization of the spectra in Figure 9 results in the frequency response curves shown in Figure 10. As illustrated, the resulting data points (open circles) for the two fluorophores Rhodamine B and Rose Bengal agree quite well with the Lorentzian curve (solid line) corresponding to the literature lifetime (3.2 and 0.9 nsec, respectively). In the case of Rose Bengal these data are observed to be skewed to higher frequencies than the literature lifetime would suggest. In fact, the true lifetime for the solution being employed in these investigations was found by an independent technique to be 0.6 rather than 0.9 nsec, because of the presence of an unexpected quencher. This shorter lifetime is in excellent agreement with the recorded data.

To obtain the impulse response function from a fluorophore using a cw laser or some other randomly modulated source, a nanosecond correlator

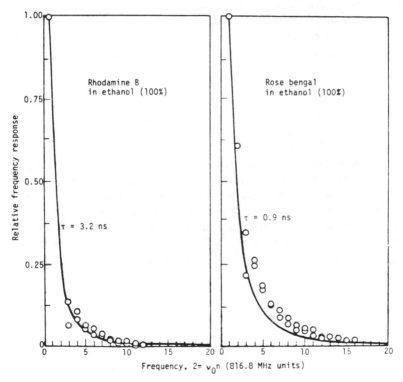

FIGURE 10. Normalized Lorentzian frequency response curves of two fluorophores, obtained from the data shown in Figure 9. Open circles denote data points; solid lines represent Lorentzian curves fit to the literature lifetime for each fluorophore (given as τ values). From Hieftje et al.,[14] reproduced by permission of the American Institute of Physics.

must be constructed. Unfortunately, such devices are not commercially available and must be constructed in the laboratory. To appreciate how such a device might be constructed, let us briefly examine the correlation process itself.

Cross-correlation can be expressed mathematically as

$$C_{1,2}(\tau) = \lim_{T \to \infty} \frac{1}{2T} \int_{-T}^{+T} f_1(t)f_2(t \pm \tau) \, dt$$

In this expression $f_1(t)$ and $f_2(t)$ are two time-dependent waveforms that are to be correlated, T is the length of each of them, and $C_{1,2}(\tau)$ is the cross-correlation function itself. Because f_2 has a parameter (τ) added to or subtracted from its time position, it has been delayed somewhat with respect to the temporal location of f_1. To cross-correlate these two waveforms, they must be multiplied, their products integrated over all time, and the result divided by the integration interval. In other words, the product must be time averaged. This time-averaged product is then displayed in terms of the displacement or delay (τ) between the two waveforms.

Instrumentally, then, one can cross-correlate two time-varying wave-forms by delaying one of them, multiplying it by its undelayed partner, and expressing the time average of this product as a function of the delay. Such an instrument is illustrated schematically in Figure 11.

To implement cross-correlation on a nanosecond time scale, both the multiplier and time delay must have nanosecond time response (GHz frequency response) characteristics. A device having these characteristics is illustrated schematically in Figure 12. In Figure 12, fluctuations in the laser are cross-correlated with those in the fluorescence it induces by directing a small fraction of the laser radiation into a fast photodetector (*PD*). The fluctuating photodetector output is then cross-correlated with the output from another fast detector (*PM*) by means of a correlator consisting of a microwave mixer serving as a high-speed multiplier and a time averager constructed from a low-pass electronic filter. Conveniently, the delay

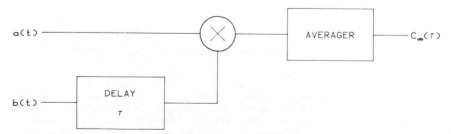

FIGURE 11. Schematic diagram of a cross-correlator. See text for discussion.

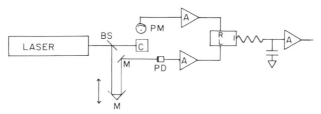

FIGURE 12. Instrument for obtaining luminescence lifetimes using a time-varying laser source and an optical cross-correlator. *PM* is the photomultiplier tube, *BS* is the beam splitter, *C* is the sample cuvette, *M* are the mirrors or reflectors, *PD* is the high-speed photodiode, *A* is the amplifier, and *R*, *L*, and *I* are reference (*R*) and local oscillator (*L*) inputs and intermediate-frequency (*I*) output of a double-balanced microwave mixer serving as a multiplier.

between the two signals can be implemented with a movable retroreflector system. Moving the reflector lengthens the path between the laser and *PD*, so that the radiation arrives at the detector a bit later. Knowing the speed of light to be approximately 3 nsec/m, one can derive an accurate time scale (3 psec/mm) by measuring the mirror displacement. The resulting cross-correlation (luminescence decay curve) can be traced out on a strip-chart recorder if the retroreflector movement is at constant velocity.

Significantly, instrumentation used to implement this approach is not much more complex than that shown in Figure 12.[16,17] Both cw[16] and mode-locked[17] lasers have been employed in this kind of application and both have yielded high-quality time-resolved decay curves. A range of fluorophores were investigated and life-time resolution down to 10 psec has already been obtained using deconvolution procedures. A similar approach, but using autocorrelation and an electronic rather than optical delay line has been applied to the measurement of picosecond laser pulses.[18]

Clearly, these new correlation-based approaches to luminescence lifetime measurements offer exciting alternatives to conventional techniques. No matter whether the frequency response or impulse response functions are measured, the perturbing radiation can be applied to the sample over extended periods of time, rather than in a single pulse. Also, when implemented using laser mode noise, no exotic mode-locking laser accessories are required. Finally, the instruments are inherently simple and inexpensive to construct and possess an inherently high degree of temporal resolution.

However, the methods exhibit several drawbacks as well. Unlike the time correlated single-photon technique, entire detector response times enter into the measured luminescence curves, requiring extensive deconvolution (the time-correlated single-photon technique is limited in time resolution by the photodetector's leading edge uncertainty rather than by its pulse width). Also, the correlation techniques inherently require a higher light flux than the single-photon method, a situation that can lead to sample photodecomposition. However, the high-speed signal processing that these new methods permit and their other advantages urge that investigation into their use be continued and that they be applied to new fluorescent systems to explore their range of applicability.

REFERENCES

1. G. G. Stokes, *Philos. Trans. R. Soc. London* **142**, 463 (1852).
2. E. Gaviola, *Z. Phys.* **35**, 748 (1926).
3. G. Horlick and G. M. Hieftje, Correlation Methods in Chemical Data Measurement, in *Contemporary Topics in Analytical and Clinical Chemistry*, Vol. 3, D. M. Hercules, G. M. Hieftje, L. R. Snyder, and M. A. Evanson, eds. (Plenum Press, New York, 1978), Chapt. 3.
4. D. Magde, E. Elson, and W. W. Webb, *Phys. Rev. Lett.* **29**, 705 (1972).
5. Y. Chen, *J. Chem. Phys.* **59**, 5810 (1973).
6. B. J. Berne and R. Pecora, *Annu. Rev. Phys. Chem.* **25**, 233 (1974).
7. B. R. Ware and W. H. Flygare, *Chem. Phys. Lett.* **12**, 81 (1971).
8. W. R. Ware, Transient Luminescence Measurements, in *Creation and Detection of the Excited State*, Vol. 1A, A. A. Lamola, ed. (Marcel Dekker, New York, 1971), Chap. 5.
9. M. A. West and G. S. Beddard, *Amer. Lab.* **8**(11), 77 (1976).
10. K. G. Spears, L. E. Cramer, and L. D. Hoffland, *Rev. Sci. Instrum.* **49**, 255 (1978).
11. R. D. Spencer and G. Weber, *Ann. N.Y. Acad. Sci.* **158**, 361 (1969).
12. E. R. Menzel and Z. D. Popovic, *Rev. Sci. Instrum.* **49**, 39 (1978).
13. A. E. W. Knight and B. K. Selinger, *Spectrochim. Acta, Part A* **27**, 1223 (1971).
14. G. M. Hieftje, G. R. Haugen, and J. M. Ramsey, *Appl. Phys. Lett.* **30** 463 (1977).
15. G. M. Hieftje, G. R. Haugen, and J. M. Ramsey, New Laser-Based Methods for the Measurement of Transient Chemical Events, in *New Applications of Lasers to Chemistry*, *ACS Symposium Series No. 85*, G. M. Hieftje, ed. (American Chemical Society, Washington, D.C., 1978).
16. C. C. Dorsey, J. M. Pelletier, and J. M. Harris, *Rev. Sci. Instrum.* **50**, 333 (1979).
17. J. M. Ramsey, G. M. Hieftje, and G. R. Haugen, *Appl. Opt.* **18**, 1913 (1979).
18. J. M. Ramsey, G. M. Hieftje, and G. R. Haugen, *Rev. Sci. Instrum.* **50**, 997 (1979).

Chapter 3

Probe Ion Techniques for Trace Analysis

John C. Wright

A. INTRODUCTION

The development of convenient and reliable tunable dye lasers offers a new and potentially important excitation source for analytical spectroscopy. The high spectral brightness of these sources permits measurement of very low concentrations down to single atoms,[1-3] while the narrow bandwidths permit a high selectivity for a particular analyte in the presence of many potential interferences. This selectivity is particularly important at very low concentration levels where there are many potential interferences. Conventional fluorescence techniques, for example, are limited by background fluorescence and not by the ability to see analyte fluorescence. However, in order to take advantage of the narrow laser bandwidths the analyte must also have narrow linewidths that are compatible with the laser's in order to achieve the high selectivity. It is the function of the analytical chemist to develop the techniques required to move the promise of laser excitation from a laboratory curiosity to a practical method for ultratrace chemical analysis. Typically, the narrow analyte linewidths are obtained by transforming the analyte to the gas phase. Of course, this transformation is a problem that the analytical chemist has a great deal of expertise in, and there is much excellent work being carried out in this approach that should prove very successful.

JOHN C. WRIGHT • Department of Chemistry, University of Wisconsin, Madison, Wisconsin 53706

There is a very different method of achieving sharp-line optical transitions that are characteristic of an analyte, which has been called SEPIL — selective excitation of probe ion luminescence. The rare-earth ions (lanthanides and actinides) are the key to the method. Rare-earth ions have sharp-line fluorescence and absorption features even in condensed phases.[4] It has been recognized for many years that the spectral lines of a rare-earth ion could be used as a spectroscopic probe of the short-range phenomenon in condensed phases because the line positions are strongly dependent upon the crystal fields generated by the immediate surroundings about the rare-earth ion.[5] Changes in the crystal field caused by impurities near the rare-earth ion can cause large changes in the spectra. However, the use of rare-earth ions as probes has not been effectively implemented because most materials of practical interest have a sufficient number of differing environments (or sites) that the task of line-sorting a spectrum could not be surmounted.[6,7] But if one has a tunable laser that can excite a specific absorption line of a rare earth in a particular environment (in general, the absorption-line positions of rare earths in different environments are sufficiently different that they do not overlap), fluorescence results only from that rare earth and the fluorescence spectra contain only that set of lines.[8] An analogous method provides a single-site excitation spectrum. By tuning a monochromator to a specific fluorescence line of a rare earth in a specific crystallographic site and scanning the dye laser over the absorption lines, one can record an excitation spectrum at the particular fluorescence wavelength selected. The combination of fluorescence and excitation spectra provides information about both the high- and low-lying electronic states of the rare-earth ions. The problem of line sorting is therefore greatly simplified and the rare earths can be used efficiently as spectroscopic probes. The narrow linewidths of the rare earths in both excitation and fluorescence furnishes a great deal of selectivity for a particular rare earth in a particular crystallographic site. The long fluorescent lifetimes of rare-earth ions (typically between $1\,\mu$sec and $10\,$msec) permits one to implement time-resolution techniques with a pulsed laser and to gain additional selectivity between rare-earth fluorescence and everything else that fluoresces. Since common interferences have much shorter lifetimes, the background that usually limits fluorescence techniques does not limit rare-earth fluorescence.

In our research we have found that the fluorescence of rare-earth ions can be measured down to very low solution concentrations and that simple precipitation procedures can be used for sample preparation.[9] Rare-earth analysis can be accomplished directly by exciting fluorescence from the rare-earth ions in the precipitates. Analysis of other ions that do not fluoresce can be done by forming associates between the analyte and a rare-earth ion.[10] The rare-earth ions that have an analyte nearby will experience crystal fields different from any of the other rare earths. By achieving a $1:1$ correspondence between the analyte and a fluorescent rare-earth ion and using the laser techniques to obtain single-site spectra from only the analyte–rare-earth associates, one should be able to determine the analyte with the same sensitivity as a rare-earth measurement alone. The key to the success of the method is achieving the association between the rare earth and the analyte. A number of different cases have been envisioned to achieve this association:

(1) Add reagents that will form a compound with the analyte ion. If small concentrations of an optically detectable rare earth are also added, the incorporation of some rare-earth ion into the lattice of the analyte compound will produce spectra characteristic of that compound.

(2) Form a lattice of a compound in the presence of small concentrations of a rare earth and an analyte ion. Random associations can be formed between the rare-earth ion and the analyte ion.

(3) Form stoichiometric, optically active rare-earth compounds in the presence of analytes so any analyte that enters the lattice must affect an optically active rare-earth ion.[11]

(4) Form a lattice of a compound for which the presence of small rare-earth and analyte ion concentrations in the lattice will require charge compensation. The compound is selected so the rare-earth and analyte ions charge-compensate each other. The ensuing Coulombic interaction results in an associated complex of rare-earth and analyte ions.

(5) Form a lattice similar to case (4) where charge compensation is required. In this case, however, the rare earth forms a complex with another ion or defect and the analyte ion forms a separate complex with another ion or defect. The dipolar interactions between the two complexes cause an association and result in one large four-member complex that contains the optically active rare-earth ion and the analyte ion.

(6) Form solution complexes between the analyte and the rare-earth ion. This case is only applicable for analytes that can act as ligands.

These six cases are the basis for this chapter. First will be presented the fundamentals of the spectroscopy and defect chemistry that underlie these methods and then there will be a discussion on the present status of work on each case.

B. SPECTROSCOPIC FUNDAMENTALS

The important electronic levels of the trivalent lanthanide ions for this application are all within the $4f^n 5s^2 5p^6$ electron configuration.[4] The

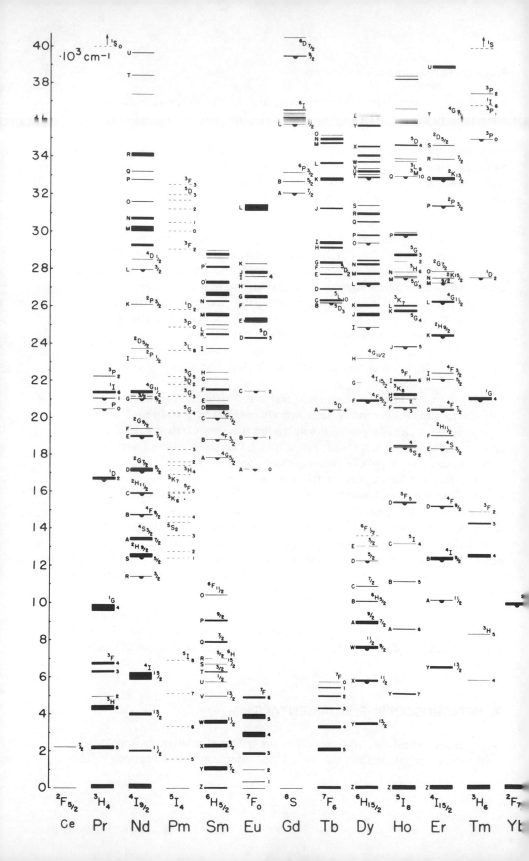

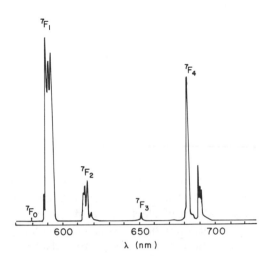

FIGURE 2. Fluorescence spectrum of $CdF_2:Eu^{3+}$ showing the $^5D_0 \rightarrow$ $^7F_{0,1,2,3,4}$ fluorescence transitions.

optical transitions result from changes within the unfilled $4f$ electron shell, which is shielded from outside chemical and electrical influences by the outer, filled $5s^2 5p^6$ orbitals. Consequently, electric and crystalline fields do not influence the electronic level positions as much as in the transition metals.[5] The main interactions that determine the level positions are the Coulombic and spin-orbit interactions.[4] The level positions for the lanthanides with an unfilled $4f^n$ shell are shown in Figure 1.[4] Although this diagram is strictly applicable for lanthanide-doped $LaCl_3$ crystals, it can be used for other crystals because of the small influence of the matrix on the level positions. One can now use this diagram to predict the wavelength region for the absorption and fluorescence transitions for each individual lanthanide. For example, Figure 2 shows the fluorescence spectrum of $CdF_2:Eu^{3+}$. From Figure 1 one can see that the lines in Figure 2 result from transitions from the Eu^{3+} $^5D_0$ state to each of the ^7F_J states, where $J = 0$ through 4. The positions of these transitions do not change markedly for other lattices.

1. Crystal-Field Splittings and Vibronic Sidebands

Although the crystal fields do not interact strongly with the $4f^n$ electron configuration, they nevertheless cause additional crystal-field splittings of each of the free ion levels (see Figure 1) into a manifold of levels that are

---

FIGURE 1. Electronic energy levels for the $4f^n$ electron configuration of the lanthanides. From Dieke[4]; reproduced by permission of Wiley.

closely spaced relative to the distances between manifolds,[4] The relative sizes of the splittings in LaCl$_3$ are indicated by the width of the levels in Figure 1. For a given level labeled by $^{2S+1}L_J$, a low-symmetry crystal field will cause it to split into $(2J + 1)/2$ levels if the lanthanide in question is a Kramers ion (an ion that has an odd number of electrons so that it has a 1/2-integer value of J, e.g., Ce^{3+}, Nd^{3+}, Sm^{3+}, Gd^{3+}, Dy^{3+}, Er^{3+}, or Yb^{3+}) or into $2J + 1$ levels if it is a non-Kramers ion (Pr^{3+}, Pm^{3+}, Eu^{3+}, Tb^{3+}, Ho^{3+}, or Tm^{3+}).[4] Thus if one looks at the transitions in Figure 2 under high resolutions such as shown in Figure 3, one sees the additional structure caused by the crystal-field splitting.

In addition to the lines observed from transitions between the individual crystal-field levels, transitions can also be observed that involve combinations of electronic and vibrational excitations.[4,12,13] These transitions are analogous to typical molecular spectra except that in a crystal field the vibrations are usually quantized according to the long-range translation symmetry of the crystal and are called phonons. If one works at very low temperatures where the phonon modes are not populated, vibronic sidebands appear on the high-energy side of pure electronic absorption transitions as absorbed photons are converted into vibrational and electronic excitations and on the low-energy side of pure electronic fluorescence transitions where an excited electronic state relaxes to a lower-electronic-state excitation and a vibrational excitation. The spacing between the pure electronic transition and its vibronic sideband is determined by the phonon energy. Generally the vibronic sidebands are much broader than the electronic transition because there are many possible phonon modes and each phonon mode has a continuum of different velocities (or, more correctly, **k** vectors) that characterized its propagation and determine its energy. The intensity of vibronic sidebands is very low, typically 500–1000 times weaker than the electronic transitions, although there are examples of much stronger transitions in favourable situations.[13] The weak interactions between the electronic levels and the lattice crystal fields result in small

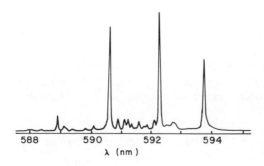

588   590   592   594
λ (nm)

FIGURE 3. Fluorescence spectrum of the $^5D_0 \rightarrow {}^7F_1$ transition of CdF_2:Eu^{3+}.

coupling with the phonon modes and the low vibronic intensities. Nevertheless, the vibronic sidebands are important in determining the ultimate selectivity of SEPIL.

2. Radiative Transition Rates

It is important to understand the factors that control the fluorescence from the levels shown in Figure 1, since only a few of these levels will actually fluoresce. Fluorescence from an excited manifold of crystal-field levels originates from the levels that are populated according to a Boltzmann distribution because the relaxation rates between levels within a manifold are much faster than the relaxation rates of levels to other manifolds. At very low temperatures only the bottom level of a manifold will be populated. There will be a characteristic radiative transition rate from a populated level to each of the levels below it. The total radiative rate (τ_{rad}^{-1}) will be the sum of all the individual radiative rates. The radiative rates for rare-earth ions are much lower than most materials because the transitions are parity forbidden, being changes within the $4f^n 5s^2 5p^6$ electron configuration. The primary mechanism that permits electric-dipole transitions is the crystal-field-induced mixing of the $4f^n 5s^2 5p^6$ configuration with other configurations of opposite parity. The resulting forced electric-dipole transition probabilities are on the order of $10^3 \sec^{-1}$ instead of the 10^8-$\sec^{-1}$ rates characteristic of more allowed transitions.[12,13] The absorption coefficients for rare-earth ions are therefore low and characterized by absorption cross sections of $\sim 10^{-19}$ cm^2.

3. Nonradiative Relaxation

Whether a given level will fluoresce depends upon how well the radiative rate competes against the nonradiative rates. There are three important mechanisms that control the nonradiative relaxation of an excited electronic level—multiphonon relaxation, energy transfer, and migration to a quenching sink.[13] Each of these mechanisms has a characteristic relaxation rate and the total nonradiative relaxation rate is the sum over all rates.

a. Multiphonon Relaxation

An excited electronic level can transfer a portion or all of its energy into lattice vibrations. This process is called multiphonon relaxation. Consider the hypothetical energy levels of ion A in Figure 4. If ion A is excited to level 3, it can relax to level 2 and give the energy difference to the lattice as phonons. The rate of multiphonon relaxation depends upon the energy gap (in this case, the difference in energy between levels 3 and 2), the energy of

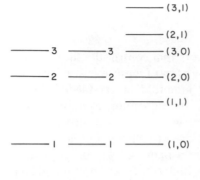

FIGURE 4. Energy-level diagram for two ions (A and B) that are near each other. The system of both ions can be described by single energy-level diagram labeled with all combinations of the energies for both ions.

the lattice phonons, the population of phonon modes, the interaction strength of the coupling between the electronic and lattice wave functions, and the character of the initial and final states involved in the relaxation.[14]

The multiphonon relaxation rate for CaF_2 is shown in Figure 5.[15] This data was obtained at 10°K, where the phonon modes are depopulated. The rate depends exponentially upon the energy gap. Thus the multiphonon rate for level 3 in Figure 4 relaxing to level 2 will be much faster than the

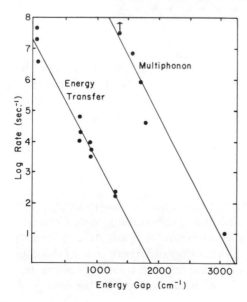

FIGURE 5. Dependence of multiphonon relaxation and nonresonant energy-transfer rates in CaF_2 on the energy gap below a level.

relaxation rates to levels 1 or 0, and the total rate therefore depends primarily upon the gap from a level to the one immediately below it.

The number of phonons created in the multiphonon relaxation is usually approximated by assuming a lattice has an "effective" phonon energy that is determined by averaging the effects of all phonons according to their importance in the relaxation.[14,16] The number of phonons is then the energy gap divided by the effective phonon energy. If one assumes an effective phonon energy of 370 cm^{-1} for CaF_2, a 2000-cm^{-1} gap would require emission of 5.4 phonons. If one is interested in a softer lattice that has lower-energy phonons (e.g., $SrCl_2$), more phonons will be required in the relaxation across the same gap and the rate will be exponentially smaller. Similarly, if the host has very-high-energy phonons, the relaxation rate is much larger. Water, for example, has vibrational energies in excess of 3500 cm^{-1} so that a 2000-cm^{-1} gap could be relaxed by a single phonon. Generally, in lattices that have groups with high energies, internal vibrations provide efficient multiphonon relaxation.

The data shown in Figure 5 are valid for very low temperatures where the phonon modes are normally not populated. At higher temperatures the occupation of a phonon mode (n) will increase according to a Bose–Einstein distribution[14,16]:

$$n = \frac{1}{e^{\hbar\omega/kT} - 1} \tag{1}$$

where $\hbar\omega$ is the energy of the phonon. Multiphonon relaxation has both spontaneous and stimulated emission components analogous to light emission. At higher temperatures, the stimulated component becomes more important according to the equation

$$W_{mp}(T) = W^0(n + 1)^N \tag{2}$$

where N is the effective number of phonons involved and mp stands for multiphonon. Figure 6 illustrates the temperature dependence of the multiphonon relaxation rate for an energy gap of ~ 2700 cm^{-1} in CaF_2.

With this information, one can now predict what levels in Figure 1 should fluoresce, at least for CaF_2. In order to compete effectively with nonradiative multiphonon relaxation, a level with a typical radiative rate of 10^3 sec^{-1} should have a gap of ~ 2400 cm^{-1} below it (see Figure 5) if the sample is at low temperatures. Referring to Figure 1, one can see that every optically active lanthanide ion should fluoresce in CaF_2 from at least one level, except for Ce^{3+}. If the lanthanide ion is in a water or ice matrix, a gap on the order of ca. 20,000 cm^{-1} is required for efficient fluorescence because

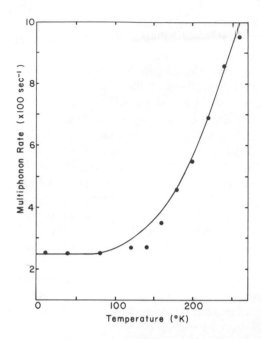

FIGURE 6. The temperature dependence of the multiphonon relaxation rate of the $^4F_{9/2}$ manifold of $CaF_2:Er^{3+}$. The solid line is a theoretical fit assuming three phonons.

of the high vibrational energies. Only Tb^{3+} and Eu^{3+} have levels with gaps that approach this requirement and, consequently, the radiative quantum efficiencies for lanthanides in aqueous solutions are low.

b. Energy Transfer

Nonradiative relaxation can also occur by energy transfer to another ion. If there is another ion in the lattice that has excited levels, an excited ion can transfer part or all of its energy to the other ion. The rate of energy transfer depends upon the distance between the ions involved in the transfer, the mechanism of energy transfer, the electronic levels of the states, and the character of the lattice and its phonons.[17] Typically, the energy transfer occurs over distances up to 50 Å because of electric-multipole interactions between ions. The radial fall-off of the energy-transfer rate depends upon the order of the multipole mechanism. Electric-dipole–electric-dipole transfer, for example, falls off as r^{-6}.[17,18] If the ions are close, exchange interactions can become important as well and increase the efficiency of the energy transfer.[19]

In order for energy transfer to be efficient, it is usually necessary for the transfer to be nearly resonant. There are numerous examples of near-resonant levels between two lanthanide ions.[7] Looking at Figure 1, the $^3F_3$ level of Pr^{3+} can transfer energy to Nd^{3+} in any of three ways.[20] The $^3F_3$ level of Pr^{3+} can relax to the ground state, exciting the Nd^{3+} ion to the $^4I_{15/2}$ level; the $^3F_3$ level can relax to the $^3H_5$ level, exciting the Nd^{3+} ion to the $^4I_{13/2}$ level; or the $^3F_3$ level can relax to the $^3H_6$ level, exciting the Nd^{3+} ion to the $^4I_{11/2}$ level. Experimentally, one observes the Pr^{3+} fluorescence quantum efficiency drop as the Nd^{3+} concentration increases to concentrations of 0.1 mol %. The fluorescence decay after a pulsed excitation becomes nonexponential when this type of energy transfer becomes important because the decay rate of any ion depends upon the distance to the nearest ion that can accept its energy. The overall decay becomes an ensemble average of the decays of each ion. More importantly, for analytical situations the energy transfer from a lanthanide can occur to transition metals that typically have broadbands which can resonantly accept the excitation energy from a lanthanide. Because the $3d^n$, $4d^n$, or $5d^n$ orbitals of transition metals are not shielded as well as the lanthanide $4f^n$ orbitals, the energy transfer can be more efficient.[13] Many of the transition metals can be effective fluorescence quenchers.

In lattices where the lanthanides require charge compensation, the situation changes markedly, because in these systems there is a marked tendency for the lanthanides to cluster and their radial separations are no longer random.[21] The short separations in the clusters promote very efficient energy transfer, while the fixed separations define a unique energy transfer rate for an individual ion.[15,22] Fluorescence decays from the ions in the clusters are then exponential, even with energy transfer. Figure 4 shows the energy levels for two ions in such a cluster on the right, while on the left the possible energy states for the combination of both ions is represented. If level 3 of ion A is initially excited, energy transfer can occur where ion A relaxes to level 2 while exciting ion B to level 1. In the double-ion energy levels shown on the right, this sequence corresponds to relaxation from double-ion level $(3, 0)$ to $(2, 1)$. In general, levels $(3, 0)$ and $(2, 1)$ are not resonant and energy must be dissipated as lattice phonons. This nonresonant energy transfer between ions within clusters depends exponentially on the energy gap between the double-ion levels (see Figure 5) in an manner analogous to multiphonon relaxation.[15] The rate is much lower than the comparable rate for multiphonon relaxation because it involves the strength of interaction between two ions as well as the coupling to the lattice. Nevertheless, the rates are quite large and make nonresonant energy transfer an important process for understanding fluorescence from clusters. We shall see later how nonresonant energy transfer can have important applications.

c. Migration to Quenching Sinks

The energy-transfer processes are all dependent upon the concentrations of the quenching ions. If a lanthanide concentration is raised to relatively high values, another nonradiative relaxation mechanism becomes important.[23] An excitation can resonantly migrate through a lattice at high dopant concentrations because there will be an identical ion nearby. The migration can continue until it reaches an ion that has a different ion, such as a transition metal, near enough to receive the excitation and the excitation is lost. The importance of this quenching mechanism depends upon the material purity. Generally migration to sinks becomes important at concentrations above 15 mol %.[23]

The total relaxation rate of any level (ω_T) will be the sum of all the radiative and nonradiative relaxation rates:

$$\tau^{-1} = \omega_T = \omega_R + \omega_V + \omega_{ET} + \omega_M \tag{3}$$

where τ is the lifetime of the level, ω_R is the radiative rate, ω_V is the multiphonon relaxation rate, ω_{ET} is the energy transfer rate, and ω_M is the rate of concentration quenching by migration to a sink.

The radiative quantum efficiency (η) becomes

$$\eta = \omega_{\text{rad}}/\tau^{-1} \tag{4}$$

4. Linewidths

The relaxation rates of levels determine the sensitivity that an analysis procedure is capable of, while the linewidths of levels determine a procedure's selectivity. The linewidths of a transition have both homogeneous and inhomogeneous components. The homogeneous width of a level depends upon its lifetime through the uncertainty principle.[24,25] The lowest level in a manifold is usually quite sharp because its lifetime is usually long. The upper states of a manifold are usually broad because they can relax efficiently ($\sim 10^{12}$ sec^{-1}) by single-phonon relaxation within a level. There is additional broadening at higher temperatures ($>20°K$) as the crystal fields acquire a time-varying component from the vibrating lattice (i.e., phonons scatter off the lanthanide).[13,25] This broadening hampers the use of these probe ion techniques at room temperature because the selectivity decreases as the lines broaden and the fluorescence intensities decrease if a narrow-band laser is used for excitation because the peak absorption must decrease as the line broadens. The best selectivity is therefore achieved by working at low temperatures and by using transitions between levels that cannot relax rapidly by single-phonon relaxation.

The inhomogeneous component to the linewidth is caused by variations in the crystal fields encountered at different places in the lattice because of strains, dislocations, and other deviations from perfect long-range orders.[13,25] The inhomogeneous width is very sensitive to the lattice perfection. Lattices that would appear to have excellent long-range order from powder x-ray diffraction measurements can still exhibit large inhomogeneous widths from the lanthanides in them.

C. SOLID-STATE CHEMISTRY FUNDAMENTALS

1. Phase Partitioning

The introduction of dopants into lattices is described by a distribution coefficient that defines the relative concentrations in the phases that contain dopants.[26] If the dopant is uniformly distributed through the phases and has reached its equilibrium distribution, the distribution coefficient is labeled D, and is defined by

$$D = f_D a_L / a_D \qquad (5)$$

where f_D is the mole fraction of dopant ion, a_L is the activity of the lattice ion replaced by the dopant ion, and a_D is the activity of the dopant in solution.

If the dopant is distributed heterogeneously because of kinetic limitations such as frequently occur in coprecipitation, the distribution coefficient is labeled λ and is defined by

$$\lambda = \log \frac{(a_D)_{t=0}}{(a_D)_{t=\infty}} \bigg/ \log \frac{(a_L)_{t=0}}{(a_L)_{t=\infty}} \qquad (6)$$

where a_D and a_L have their previous meaning, while the subscript indicates initial or final concentrations. This relation is the well-known Doerner–Hoskins equation. λ becomes identical to D for the deposition of an infinitesimal layer in the formation of a lattice or for very slow rates of lattice formation, where internal diffusion can become important. The size of λ or D depends upon the difference in free energy when the dopant is in each of the two phases.[27]

The value of D or λ will only be unique when the free energy of the dopant in the lattice is a constant. That can be changed markedly in situations requiring charge compensation (as we shall see) because of the solid-state chemistry of defects in the lattice. It is important to understand how the defects affect the dopants because of the close relationship to what is observed spectroscopically.

2. Defect Equilibria

There are three types of point defects that occur in materials of interest here.[28-31] *Frenkel defects* occur when a lattice cation leaves its normal lattice position and lodges itself in an interstitial position, creating a cation vacancy and a cation interstitial. *Anti-Frenkel defects* occur when a lattice anion leaves its normal lattice position and lodges itself in an interstitial position, creating an anion vacancy and an anion interstitial. Finally, *Schottky defects* occur when an anion and a cation leave their normal lattice positions and move to the surface of the sample, enlarging the volume and creating an anion–cation-vacancy pair. The concentration of defects will be temperature dependent, but at any finite temperature one must have some defects. The defect concentration is also dependent upon the surrounding atmosphere. If the atmosphere has a high vapor pressure of the cation because of its deliberate introduction, the concentration of cation vacancies can be diminished. Similar behavior can occur for anions that might be in the atmosphere. One can consider the formation of the intrinsic defects as the dissociation of the perfect lattice (the solvent), in analogy to the formation of the "water defects" OH^- and H^+ by the dissociation of H_2O.

If a cation dopant replaces a lattice cation that has a different charge, the lattice must provide some charge compensation. The intrinsic charge compensation depends upon which of the three defect mechanisms is important. In CaF_2, for example, the defects are anti-Frenkel fluoride vacancies ($V_F^·$) and fluoride interstitials (F_i').[32] The notation of Kröger and Vink is being used to designate the defect species.[32] The main symbol designates the moiety that is at the lattice position indicated by the subscript. The net charge of the species relative to the normal lattice is indicated by a superscript dot (·) if it has an extra positive charge, a prime (′) if it has an extra negative charge, and an x if it has no extra charges. If the trivalent lanthanide Er^{3+} substitutes for a divalent Ca^{2+}, an extra fluoride interstitial provides the charge compensation.[28,32] There will be a Coulombic attraction between the extra negative charge of the fluoride interstitial and the extra positive charge of the erbium dopant, which will encourage associated pairs to form—$(Er_{Ca}·F_i)^x$. If Na^+ were added to CaF_2, one would expect it to be charge compensated by fluoride vacancies, which could form $(Na_{Ca}·V_F)^x$ associates.[32] There will also be dipolar interactions between pairs, which would encourage the formation of clusters of the pairs. In addition, the Na_{Ca}' and $Er_{Ca}^·$ have opposite charges relative to the lattice and will also tend to form pairs. Of course, these pairs could also form clusters with each other or with other pairs. The majority of this information can be summarized by writing the following series of equilibria:

$$
\begin{array}{ccc}
\text{Perfect CaF}_2 \text{ lattice} \rightleftharpoons F_i' & + & V_F^{\cdot} \\
+ & & + \\
Er_{Ca}^{\cdot} & + & Na_{Ca}' \quad \rightleftharpoons (Er_{Ca}\cdot Na_{Ca})^x \\
\updownarrow & & \updownarrow \\
\cdots + (2Er_{Ca}\cdot 2F_i)^x \rightleftharpoons (Er_{Ca}\cdot F_i)' + (Er_{Ca}\cdot F_i)^x & (Na_{Ca}\cdot V_F)^x
\end{array}
$$

$$(7)$$

These equilibria are analogous to the familiar EDTA equilibria:

$$
\begin{array}{ccc}
H_2O \rightleftharpoons OH^- & + H^+ & \\
+ & + & \\
Mg^{2+} & + Y^{4-} & \rightleftharpoons MgY^{2-} \\
\updownarrow & \updownarrow & \\
Mg(OH)^+ & HY^{3-} & \\
+ & + & \\
\vdots & \vdots &
\end{array}
$$

$$(8)$$

where it is well known that pH must be controlled for analytical usage of EDTA. By analogy, the intrinsic anti-Frenkel defect concentrations must be controlled if the lanthanides are to be used analytically. The potential analytical utility in Equation (7) arises from the $(Er_{Ca}\cdot Na_{Ca})^x$ moiety. The crystal-field levels of the Er^{3+} ion will reflect the presence of a Na^+ ion in its vicinity, and by selectively exciting Er^{3+} ions in such environments, one can identify and quantify the presence of Na^+. The following examines the consequences of the coupled-defect equilibria, using the CaF_2 case as an example.

There are a number of simplified extreme cases for the equilibria that are listed below as particularly interesting:

(1) A low formation constant K_f for $(Er_{Ca}\cdot F_i)^x$ encourages dissociation of $(Er_{Ca}\cdot F_i)^x$, especially at low Er^{3+} concentrations; it also encourages dissociation of $(Na_{Ca}\cdot V_F)^x$ and association of $(Er_{Ca}\cdot Na_{Ca})^x$.

(2) A low K_f for $(Er_{Ca}\cdot Na_{Ca})^x$ encourages dissociation of $(Er_{Ca}\cdot Na_{Ca})^x$, especially at low Er^{3+} or Na^+ concentrations; it also encourages the association of $(Er_{Ca}\cdot F_i)'$ and $(Na_{Ca}\cdot V_F)^x$.

(3) A high K_f for $(Er_{Ca}\cdot Na_{Ca})^x$ encourages association of $(Er_{Ca}\cdot Na_{Ca})^x$ and dissociation of $(Er_{Ca}\cdot F_i)^x$ and $(Na_{Ca}\cdot V_F)^x$.

(4) A high K_f for $(Na_{Ca}\cdot V_F)^x$ encourages dissociation of $(Er_{Ca}\cdot Na_{Ca})^x$ and association of $(Er_{Ca}\cdot F_i)^x$.

(5) A high K_f for $(\text{Er}_{\text{Ca}} \cdot \text{F}_i)^x$ encourages dissociation of $(\text{Er}_{\text{Ca}} \cdot \text{Na}_{\text{Ca}})^x$ and association of $(\text{Na}_{\text{Ca}} \cdot \text{V}_\text{F})^a$.

In case (1), for example, a low formation constant for the $(\text{Er}_{\text{Ca}} \cdot \text{F}_i)^x$ pair will result in dissociation of the pair, which increases the concentration of F_i'. This increase must produce a decrease in the concentration of V_F' which causes the $(\text{Na}_{\text{Ca}} \cdot \text{V}_\text{F})^x$ pairs to dissociate to bring the V_F' higher. The Na_{Ca} released by this dissociation can then combine with the higher Er_{Ca} concentration to form a higher concentration of $(\text{Er}_{\text{Ca}} \cdot \text{Na}_{\text{Ca}})^x$. The analytical consequences will depend upon which species is the one of interest. If one is performing an analysis for Er^{3+}, either the $(\text{Er}_{\text{Ca}} \cdot \text{F}_i)^x$ or Er_{Ca} would commonly be used and therefore the position of the equilibrium involving them should not be a function of other concentrations such as Na^+ or Er^{3+}. Similarly, if one is performing an analysis for Na^+, the $(\text{Er}_{\text{Ca}} \cdot \text{Na}_{\text{Ca}})^x$ species would be used and the position of the equilibrium that forms it should be independent of all concentrations. This situation is not realized in practice unless the formation constants for the species are so large that the equilibria cannot be appreciably disturbed or unless buffers are used to maintain the position of the equilibria. The size of the formation constants depends upon the free energy for the associates relative to the dissociated components. The free energies depend upon the size of the stabilization energies of the Coulombic attraction of the individual components and the lattice strain energies associated with the paired and unpaired components.

These equilibria can have a profound effect upon the distribution coefficient. Consider the distribution coefficient for Na^+ coprecipitated in CaF_2. If Er^{3+} is also present, the formation of $(\text{Er}_{\text{Ca}} \cdot \text{Na}_{\text{Ca}})^x$ pairs can make the incorporation of Na^+ much more favorable if the free energy for forming such pairs is favorable. The distribution coefficient for Na^+ would then be a strong function of the Er^{3+} concentration.

D. EXPERIMENTAL METHODS

1. Sample Preparation

To use probe ion techniques in solids, two basic preparation steps are required: formation of the solid host lattice and incorporation of dopant ions. There are four methods that have been used to form the lattice. One of the simplest is to precipitate the desired lattice from solution. We have found that the spectral transitions from lanthanides in precipitates are broad, many times too broad to permit identification of individual crystal-field transitions. The short-range order in precipitates is simply too low to permit their direct use, even if the precipitates have been aged for a long time. In order to perfect the lattice, the precipitates must be heated to temperatures where the ions become mobile. One is then restricted to precipitates that do not decompose before their lattices perfect. Precipitation has several advantages: It is a simple way of making the lattice, the probe and analyte ions can often be easily introduced into the lattice by coprecipitation, precipitation serves as a preconcentration step (especially if the coprecipitation is favorable), and it serves as a separation step because many ions that are potential interferences will be prevented from entering the lattice because of size or charge mismatches. One can also form the lattice by conventional crystallization techniques, either evaporation of the solvent or cooling. The solvent can be a standard aqueous solution or it can be a flux that must be held at high temperatures. It is not necessary to take the same time and care required to obtain large crystals by these methods because the lanthanide probe ions sense only the short-range order. One can easily use the microscopic crystals in the powders that result from rapid evaporation or cooling. A third method is to use a solid-state reaction. The reactants are finely ground together so that they are in intimate contact and are reacted at high temperatures where they can freely diffuse to form the product lattice. It is common in these sintering reactions to use reactants that will yield only the product lattice while other constituents vaporize. In the last sample preparation method, the sample can be heated above its melting point and then cooled to form the lattice.

The dopants can be introduced in several ways as well. If the host lattice is precipitated from solution, the dopants can be coprecipitated. This method has the advantage that it acts as a preconcentration step and it also provides a separation of the probe and analyte ions from contaminants that may be present. The probe and analyte ions can also be added as a separate phase. Solutions of them can be dripped on the powder of the host lattice or the reactants that will make the lattice, heated to evaporate the solvent, and ignited to a temperature where they can diffuse into the lattice. Alternatively, the ions can be added as powders to the host lattice or reactants, ground together, and ignited.

The incorporation of the ions in the lattice can be an equilibrium situation or a kinetically controlled one. The former is preferred for analytical work because it would have better reproducibility. If the situation is kinetically controlled, it becomes very important to control the temperature, the rates of heating and cooling, and all of the factors that affect the diffusion of ions in the lattice (including the intimacy of the mixing, the perfection of the lattice, and the influence that other foreign ions can have on the diffusion coefficient).

All of these methods introduce a number of experimental variables that can affect the results in profound ways. All of the parameters associated with the ignition must be optimized. The temperature must be high enough to allow ions to diffuse, but not so high that the analyte–probe ion associates break up. In many materials, different phases will be encountered as the temperature is raised and the properties of each phase differ. The time of ignition must be long enough to reach a final equilibrium or it must be adjusted for optimal results in a kinetically controlled situation. The cooling rate is important because at some point the temperature will be too low for ions to diffuse any more and the defect equilibria will be frozen in the lattice at that point. If the temperature is lowered too quickly, the defect equilibria will be frozen in at a higher "effective temperature," which favors dissociation of analyte–probe ion associates.

The atmosphere surrounding the sample during the ignition will influence the defect equilibria. It is most convenient to ignite in air but this restricts one to oxidizing conditions. The partial pressure of gases that can enter or leave the sample can determine the positions of defect equilibria. For example, the concentration of oxide vacancies in oxide hosts will depend upon the partial pressure of oxygen in the surrounding atmosphere. As seen earlier, the vacancy concentration is analogous to pH in controlling the equilibria.

Besides having an atmosphere around the sample, it is also common to have a crucible. Typical crucibles include platinum, quartz, Vycor, alumina, and porcelain. At high temperatures, the crucible material has a finite mobility and can enter the sample. In addition, the crucible must be kept clean to ensure that other materials are not available to enter the sample.

There are numerous important ways of preparing individual samples. The choice of the initial reactants can affect the defect equilibria. Samples of CaF_2 prepared by precipitation and ignition of NH_4F and $Ca(NO_3)_2$ behave differently from those prepared with NaF and $Ca(NO_3)_2$, where the Na^+ can enter the lattice.[9] Ions other than the analyte and probe ions can enter the lattice either unintentionally, as interferences, or intentionally, as solid-state defect buffers, and alter the position of defect equilibria. Grinding is sometimes necessary to provide a uniform composition of sample and a large surface area for efficient sintering. The grinding step, though, can introduce dislocations and strain fields in the lattice. It can also be a source of impurities. Leaching the sample with strong acids, bases, complexing agents, or solvents can be useful in removing unwanted phases, particularly after an ignition.

2. Selection of Materials

The key to successful analysis by SEPIL lies in the choice of the material for the host and its proper preparation. There are a number of criteria that determine whether a material is suitable:

(1) The material must be able to incorporate the probe and analyte ions into the lattice structure. This requirement implies that in most cases the analyte and probe ions substitute for lattice ions and they can therefore not differ too greatly in size from the ions they replace.

(2) The material must not quench fluorescence from the probe ion.

(3) The lattice must have a high enough short-range order for the spectral transitions to be narrower than the separations between crystal-field transitions. We have observed that materials prepared by precipitation, solid-state sintering reactions, or with fluxes must be heated to temperatures where ions may diffuse in order to achieve enough order. Materials crystallized from solution often already have a high enough short-range order, even if the solvent is driven off rapidly. If the lines are broad, there is a loss not only of selectivity but also of signal because the absorption coefficient at the peak of a transition must decrease if the transition broadens. Low short-range order can also cause the appearance of a continuum of sites that produce a broad background fluorescence which can obscure transitions from low-concentration sites.

(4) The probe and analyte ions must be associated in the lattice in positions sufficiently close that the crystal-field splittings of the probe ion are changed by the presence of the analyte ion.

(5) The site distribution should not be dependent upon the concentration of analyte or probe ions; otherwise the working curve will become nonlinear.

(6) The site distribution should be insensitive to the presence of other impurities that might be present in the lattice. If impurities can change the distribution, they will interfere with the measurement.

(7) The probe ion should have a large absorption coefficient and favorable radiative quantum efficiency in the material.

The next step is to be able to experimentally determine how well a particular material meets the requirements. There has been very little work done in this area by previous workers and there is no way of determining beforehand whether a material or preparation method will or will not be suitable. As mentioned before, there are many variables that can enter in the sample preparation steps. One can also choose any of the lanthanide ions as a probe ion and one does not know which ions can be associated with the probe ion in the particular host lattice. The uncertainty associated with each required choice leads to a plethora of different possibilities that make experiments in identifying suitable materials difficult.

In order to test whether a material is suitable we have generally compared spectra from samples that have large concentrations of probe and analyte ions (typically between 10^{-3} and 10^{-1} mol % of the host-lattice concentration) with the spectra from samples that have no analyte ions. New lines that appear in the spectra indicate a newly formed site that is unique to an associated probe–analyte ion pair. The new lines are, however, only an indication and cannot be taken as proof of an association. The presence of analyte ions could also change the site distribution of intrinsic sites without actually pairing with the probe ion. For example, if Na^+ had a weak association with fluoride vacancies in Equation (7) as well as a weak association with Er^{3+}, the fluoride vacancy ($V_F^{\cdot}$) concentration would be larger and would suppress the fluoride interstitial (F_i') concentration. The $(Er_{Ca} \cdot F_i)^x$ pairs would be encouraged to dissociate, changing the site distribution and causing the appearance of new lines characteristic of the dissociated pair.* The presence of analyte ions could also lead to the formation of a different lattice phase or could change the kinetics of forming different intrinsic sites, thus leading to new lines. None of these situations, though, leads to lines characteristic of the analyte, since other similar ions can have the same effect. It is difficult to distinguish between situations where the appearance of new lines is specific for the analyte and where it is not. One can only look for the presence of the lines in situations where the analyte is absent. If they do appear in other situations, the lines are not specific for the analyte.

In a material that has not been previously studied, it is time consuming to identify where all of the excitation and emission transitions occur and to sort the spectral transitions into those that have common site origins. Much of this work can be simplified or eliminated by using Eu^{3+} as the probe ion. The $^7F_0$ ground state and the $^5D_0$ excited state of Eu^{3+} (see Figure 1) are both singlet states ($J = 0$, which can have $2J + 1 = 1$ degenerate states) and cannot split in a crystal field. They can be shifted, though, by second-order crystal-field interactions that mix higher J states with them. If one monitors the fluorescence from the $^5D_0$ state to a lower ^7F_J state with an instrument that has a very broad bandpass which can detect fluorescence from all crystallographic sites and then scans a dye laser across the region of the $^7F_0 \rightarrow {}^5D_0$ excitation transition, each excitation line observed must correspond to a Eu^{3+} occupying a different site. This excitation scan will therefore contain information about how many sites are present and how many ions are in each site in the sample (at least for the sites that fluoresce). The intensity of the lines cannot be used to obtain the absolute concentration, because one does not know the radiative transition probabilities, but it can be used to follow how a particular site changes in concentration when the sample preparation conditions are altered. This method provides a fast

---

* It should be noted that this discussion uses Equation (7) as an example only and does not correspond to the actual behavior of $CaF_2 : Er^{3+}$, Na^+.

and simple way of surveying the behavior of sites in a material. However, there are some difficulties associated with this procedure. Since the shifts in the $^7F_0 \to {}^5D_0$ transition depend on second-order interactions,[13] they are not highly sensitive to the alterations in the crystal field caused by analyte ions. Since there can only be one transition per site, it is also possible for two sites to accidently have the same line position and prevent one from seeing that there were two sites. Finally, the $^7F_0 \to {}^5D_0$ transition is forbidden by a selection rule against $J = 0$-to-$J = 0$ transitions and only becomes allowed because of J mixing by the crystal field.[12] The intensities are therefore lower than other transitions. It is also possible that Eu^{3+} could be reduced to Eu^{2+} and thus obscure some important information about the material.

Having obtained information on how many sites are present from the $^7F_0 \to {}^5D_0$ excitation spectrum, one can tune a narrow-bandpass monochromator to the now known wavelength of each $^5D_0 \to {}^7F_0$ fluorescence transition (it will be at the same wavelength as the excitation transition because the Stokes shift for lanthanides is very small) and scan the dye laser across another excitation region, for example, the $^7F_0 \to {}^5D_2$ transition region (see Figure 1). If the line monitored is unique to a specific site, that excitation spectrum should have at most $2J + 1$ transitions or, in our example, 5. This procedure will eliminate many of the previous problems of accidentally overlapping wavelengths and low intensities, but it increases the number of transitions that will be observed for any one site. Once a suitable host and sample preparation method is obtained, one is free to select any of the lanthanides and any of their excitation or fluorescence transitions to optimize the signal levels and to minimize any spectral interferences.

3. Instrumentation

A block diagram of the experimental apparatus is shown in Figure 7. A 400-kW-peak-power nitrogen laser is used to excite a transversely pumped dye laser of the traditional Hänsch design.[33] Although the optical components in commercial dye lasers are quite expensive, we have found that much cheaper optics give essentially the same performance. An echelle grating (Model TF-R2, PTR, Inc., Ann Arbor, Michigan) with a 63°35′ blaze angle and 316 grooves/mm is used in Littrow configuration in a high order between 8th and 15th. This grating provides the dispersion characteristic of a high-angle grating, but its free spectral range is not so limited that dyes can lase in two different orders simultaneously. The grating is scanned with a stepper-motor-driven sine bar, so wavelength scans are linear. The telescope in the laser cavity consists of a simple 5-mm-focal-length diverging lens and a 120-mm-focal-length achromatic doublet (LAU 117, Melles Griot, Irvine, California). Both lenses are inexpensive, yet the effective focal

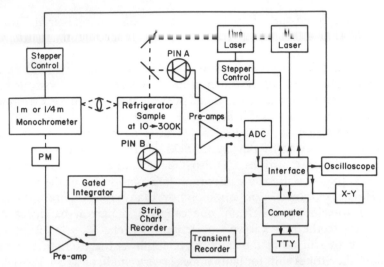

FIGURE 7. Block diagram of spectroscopic system.

length of the telescope does not depend appreciably on wavelength and permits one to scan the laser wavelength over the bandwidth of typical dyes without the need for refocusing. The laser bandwidth that can be achieved with this telescope and grating varies from ca. 0.008 to 0.02 nm, depending upon the order and dye used. A small cuvette (ESCO Scientific, Oak Ridge, New Jersey) with sides wedged at 1.5° to prevent parasitic oscillations is used as a dye cell. A small stirring bar is used to avoid thermal effects that would distort the optical path. Dye solutions can be easily interchanged with this arrangement and the 3-ml total volume lowers the consumption rate of expensive dye solutions. The output mirror is a simple, uncoated, wedged piece of glass (Edmund Scientific, Barrington, New Jersey) that provides the high-output coupling required for high-gain lasers and prevents etalon effects that would occur for optics with parallel sides.

The dye laser beam is focused onto the surface of a sample as a thin, vertical line using a spherical and cylindrical lens. The samples are usually powders that have been pressed into depressions in a copper block. The copper block is screwed onto the bottom of a cryogenic refrigerator (CSW-202, Air Products and Chemicals, Inc., Allentown, Pennsylvania) that cools the samples from room temperature to 10°K or any temperature in between. The low temperature is valuable in narrowing the lines, reducing the number of observable transitions, and decreasing multiphonon relaxation rates.

The fluorescence is measured with either a 1/4-m (Jarrell-Ash, Waltham, Massachusetts) or a 1-m (Interactive Technology, Los Gatos,

California) monochromator, depending upon the required bandpass. If one is interested in observing fluorescence from all sites, the 1/4-m instrument is used, whereas if one is interested in observing specific fluorescence lines from specific sites, the 1-m instrument is used. A photomultiplier (9658R, EMI Gencom Division, Plainview, New York) or an intensified silicon vidicon (1205D vidicon with 1205A OMA console, Princeton Applied Research, Princeton, New Jersey) can be used for detection. The output from the photomultiplier can be measured on either a home-built gated integrator, a photon counter (Ortec, Oak Ridge, Tennessee), or a transient recorder (802, Biomation, Cupertino, California). The results can be displayed on a chart recorder or can be signal averaged or processed in a small computer (PDP8/F, Digital Equipment Corp., Maynard, Massachusetts) that can control the wavelength axes of the dye laser or monochromator. A portion of the laser beam is diverted on each pulse to a spectrally corrected photodiode (PIN 10DF, United Detector Technology, Santa Monica, California) that can be used to ratio the fluorescence intensity to the excitation intensity with analog ratio electronics or with the computer. The entire system is mounted on a set of optical rails that were pioneered by J. P. Walters in order to achieve convenient and reproducible alignment of the optical system.[34]

E. EXPERIMENTAL RESULTS

1. Lanthanide Analysis

The development of a lanthanide analysis technique has been a cornerstone to the probe ion methods in providing direction for research in the methods of achieving probe-analyte ion association and a standard by which to measure the potential of ideas. These studies have centered around using CaF_2 as a host matrix for lanthanide ions. The lanthanide analysis is performed by adding $Ca(NO_3)_2$ to the solution containing the lanthanide to be analyzed and precipitating 90% of the Ca^{2+} by adding NH_4F. The lanthanides coprecipitate in the CaF_2 very favorably. Typically 96% of all the lanthanides in solution have coprecipitated after only 10% of the Ca^{2+} has precipitated. The precipitates are aged, filtered, washed, dried and ignited in a high-temperature furnace. The powder is then pressed into the sample holder and studied.

The excitation spectrum for an unignited CaF_2 precipitate with Er^{3+} coprecipitated is shown in Figure 8a. This spectrum was obtained by monitoring the green fluorescence between the $^4S_{3/2} \rightarrow {}^4I_{15/2}$ levels with the low-resolution 1/4-m monochromator (so fluorescence from all the sites

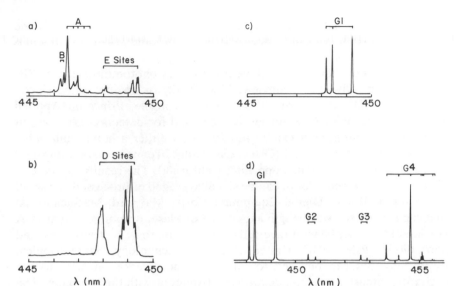

FIGURE 8. Excitation spectrum of the $^4I_{15/2} \rightarrow {}^4F_{5/2}$ transition of $CaF_2:Er^{3+}$ precipitate: (a) unignited; (b) 350°C; (c) 500°C; (d) 1000°C.

could be observed) while scanning the wavelength of the dye laser across the region of $^4I_{15/2} \rightarrow {}^4F_{5/2}$ absorption.[9] Assuming that only the bottom level of the $^4I_{15/2}$ ground state is populated at low temperatures, one would expect $(2J + 1)/2$, or 3 lines for this particular excitation spectrum, if the lines come from only one type of site. There are two sites, labeled A and B, that predominantly contribute to the spectrum.[8] Site A has been shown to consist of an Er^{3+} ion surrounded by eight fluoride ions and an additional fluoride in the nearest-neighbor interstitial position to act as the charge compensation. Site B has been shown to consist of an Er^{3+} ion with eight fluoride neighbors and a fluoride in the next-neighbor position. The site where only eight fluorides surround the Er^{3+} ion and where the fluoride interstitial compensation is very distant is allowed in these samples, but it cannot be observed optically because it has a cubic symmetry that forbids electric-dipole transitions.[8] There are other lines in Figure 8a that are not labeled and correspond to sites that are observed only in precipitates. Their character is unknown.

If one constructs a calibration curve for Er^{3+} concentration in the original solution using any of the lines in Figure 8a, one finds that the curve is very nonlinear. It plummets rapidly to zero intensity as the Er^{3+} concentration is lowered. One can only detect 5×10^{-3} mol % Er^{3+} in the CaF_2 precipitate, which corresponds to $1.25 \mu g/ml$ in the original solution under the conditions used. It is believed that the nonlinear behavior is caused by changes in the site distribution as the Er^{3+} concentration changes. At low concentrations the sites that are associates of an Er^{3+} and a fluoride interstitial dissociate in analogy to complexation equilibria where low concentrations of a loosely bound complex favor dissociation. The nonlinear calibration curve and the high detection limit sharply limit the use of unignited CaF_2 precipitates for analytical work.

The spectra of $CaF_2:Er^{3+}$ precipitates change markedly if they are heated, as shown in Figures 8b, 8c, and 8d. The A and B site lines disappear and are replaced by lines that have been shown to be characteristic of Er^{3+} clusters (Figure 8b). The single Er^{3+} ions with their fluoride compensation can associate with each other to form complex clusters that can themselves distribute in the lattice with a long-range order to form a lattice with superstructure. At higher temperatures (Figure 8c) the lines characteristic of these clusters disappear and are replaced by lines from a site that has been labeled $G1$. This site has been shown to correspond with one where an Er^{3+} ion is surrounded by seven fluoride ions and one oxide ion in the normal fluoride lattice positions.[9,35] The interstitial fluoride that served as charge compensation at lower temperatures has thus been replaced by oxygen compensation. The conversion probably involves the reaction of water trapped within the precipitates with fluoride interstitials to form HF and oxide ions that remain in the lattice.[9] There is a range of several hundred degrees over which $G1$ is the only prominent site. The lines from this site are sharp and intense and are well suited for analytical work.

If the precipitates are heated to a higher temperature, additional lines appear from sites that have more oxide ions incorporated.[9] These sites are labeled $G2$, $G3$, and $G4$ (see Figure 8d). The $G4$ site has been proposed to consist of an Er^{3+} ion surrounded by four oxide ions in a tetrahedral arrangement and a single fluoride ion, all occupying positions of normal lattice fluorides.[35] The three remaining lattice positions are not occupied in this model. It is tempting to assign $G2$ and $G3$ to the oxygen-compensated sites that are intermediate between $G1$, which has one oxide ion in its vicinity, and $G4$, which has four oxide ions, since the temperature dependence of the intensity of these sites is intermediate between $G1$ and $G4$. There is no conclusive evidence that this is the case though.

If one continues to heat the CaF_2 precipitate for an extended period of time, the spectrum continues to change as three lines from a site labeled H become important. We have shown that these lines are the same as those observed in CaO and therefore the immediate vicinity about the Er^{3+} has been changed to the CaO structure by the extended heating in air.[9] Thus there is no temperature at which the spectrum has reached a limit, a situation that is unfavorable from an analytical viewpoint. It continues to change until the phase is actually different. The kinetics associated with the changes therefore become important in establishing the reproducibility that will be observed. Of all the possible lines, those from the $G1$ site are the most attractive ones for analytical use because the $G1$ site dominates over a wide range of temperatures. Since it is the only site that appears over this range, the intensity will not be sensitive to factors that change the site distribution.

If one monitors the intensity of the $G1$ site for a series of samples prepared in the same way at 700°C, the reproducibility one obtains is typically 8%.[9] If one uses a second lanthanide ion as an internal standard, the reproducibility improves to ca. 5%. This reproducibility is limited in part by changes in the laser intensity and wavelength that occur during the course of the measurement. The $G1$ site intensity is shown in Figure 9 as a function of concentration. The intensity begins to bend over at concentrations of ca. 0.01 mol % and reaches a limiting value.

The reason for the departure from linearity is caused by clustering of several Er^{3+} ions and their oxide compensations.[36] The new site that results from the clustering can be seen in the excitation spectrum in Figure 10. The three lines of the $G1$ site that are normally observed at lower concentrations are marked in the figure. There are a total of six lines that are associated with this site and none are thermally enhanced. As noted previously, the maximum number of lines that one can have for a $J = 5/2$ level of a single ion is

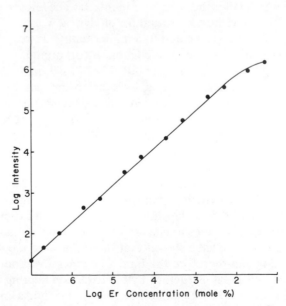

FIGURE 9. Calibration curve for the $G1$ site intensity in $CaF_2 : Er^{3+}$ precipitates.

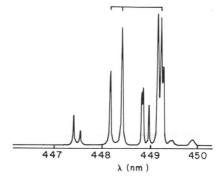

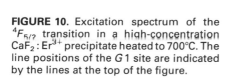

FIGURE 10. Excitation spectrum of the $^4F_{5/2}$ transition in a high-concentration CaF_2 : Er^{3+} precipitate heated to 700°C. The line positions of the $G1$ site are indicated by the lines at the top of the figure.

three. Clearly more than one ion is responsible for the lines observed. Our interpretation of the six lines is that they arise from a dimer of Er^{3+} ions and that the ground state $^4I_{15/2}$ manifold of each of the ions in the dimer is split by the magnetic and exchange interactions between the ions, thus doubling the number of transitions observed. The dimer site grows more rapidly with increasing concentration than the sites that have single Er^{3+} ions and causes the nonlinearity in the working curve when its concentration becomes comparable to the $G1$ site.

It is important to determine whether the other lanthanide ions can be analyzed by the same method. The lanthanides must be able to fluoresce efficiently from one of their excited levels. As seen previously, efficient fluorescence can be expected in CaF_2 if the gap below the fluorescing level is greater than ca. 2400 cm^{-1}. We can refer back to Figure 1 and identify which levels of the $4f^n$ configuration are likely to fluoresce efficiently. Ce^{3+} has only one such level and it is closer than 2400 cm^{-1} to the ground state. Even if it did fluoresce, the fluorescence would be in the infrared, where detection would be very inefficient. Pr^{3+} should fluoresce efficiently from the $^3P_0$ and $^1D_2$ manifolds. Nd^{3+} has only the $^4F_{3/2}$ manifold that could fluoresce, and that fluorescence lies in the near infrared, where detection efficiency is lower. Pm^{3+} should also fluoresce efficiently from the $^3F_1$ level, but Pm^{3+} is a radioactive ion that has no stable isotopes. The remaining nine lanthanides in Figure 1 have at least one suitable fluorescent level that emits in the visible or ultraviolet and can be detected efficiently. Gd^{3+}, however, requires an ultraviolet excitation that can only be provided by frequency doubling if nitrogen-pumped dye lasers are used. Consequently, we have been unable to measure Gd^{3+}. Similarly, Yb^{3+} could not be measured, because its sole $4f^n$ level lies in the infrared, where the nitrogen-pumped dye laser could not function. Table I lists the ions we have measured, the transitions used, the lifetimes of the fluorescence, and the detection limits achieved.[36] Both the Eu^{3+} and Er^{3+} detection limits were determined by contamination.

TABLE I. Analytical Parameters for Optically Detectable Lanthanide Ions

Ion	Excitation	Fluorescence	Fluorescence lifetime (msec)	Detection limit (pg/ml)
Pr	$^3H_4 \rightarrow {}^3P_0$	$^3P_0 \rightarrow {}^3H_4$	0.034	2.4
Nd	$^4I_{9/2} \rightarrow {}^2G_{7/2}, {}^4G_{5/2}$	$^4F_{3/2} \rightarrow {}^4I_{9/2}$	2.58	450
Sm	$^6H_{5/2} \rightarrow {}^4I_{9/2}$	$^4G_{5/2} \rightarrow {}^6H_{7/2}$	7.37	63
Eu	$^7F_0 \rightarrow {}^5D_2$	$^5D_0 \rightarrow {}^7F_0$	1.81	0.4[a]
Tb	$^7F_6 \rightarrow {}^5D_4$	$^5D_4 \rightarrow {}^7F_5$	3.83	38
Dy	$^6H_{15/2} \rightarrow {}^4I_{15/2}$	$^4F_{9/2} \rightarrow {}^6H_{13/2}$	2.05	15
Ho	$^5I_8 \rightarrow {}^5G_6$	$^5S_2 \rightarrow {}^5I_8$	0.017	11
Er	$^4I_{15/2} \rightarrow {}^4F_{7/2}$	$^4S_{3/2} \rightarrow {}^4I_{15/2}$	0.12	0.6[a]
Tm	$^3H_6 \rightarrow {}^1G_4$	$^1G_4 \rightarrow {}^3H_6$	Not measured	2

[a] Limited by contamination.

The detection limits are, of course, unique to the particular system used and there is always the question of how they will change if the system is changed. The detection limits listed were not determined by broadband fluorescence from impurities but by the laser power and the system noise. The long lifetimes of lanthanide fluorescence make time discrimination a very effective method for eliminating background fluorescence from other sources. The narrow excitation and fluorescence lines discriminate against interfering species with broadband spectra. Furthermore, the fluorescence is not saturated and increases in laser power will result in a corresponding increase in signal. Thus the detection limits should scale as the laser power. Typical pulse energies from our system were 0.1–0.3 mJ at a 15-Hz repetition rate. Laser are presently available that easily provide 300 times these energies and, consequently, the detection limits should be lowered considerably. The noise in our system was primarily dark noise that could be reduced to negligible proportions by cooling. This improvement should also improve detection limits.

The detection limits quoted in Table I are also dependent upon the precipitation conditions. The most natural way of expressing the detection limit is in terms of the mole percentage of lanthanide incorporation, because this is independent of the original volume of the solutions. Typically, 470 mg of CaF_2 was precipitated from an initial volume of 40 ml, but only 10 mg was used in the sample holder. The preconcentration of lanthanides can therefore be improved by precipitating less CaF_2 from a larger volume of water, thus improving the detection limits by a corresponding amount. The 0.4-pg/ml value listed for Eu^{3+} corresponds to 1.6×10^{-9} mol % in CaF_2. If one precipitated only 47 mg of CaF_2 from 80 ml, the 1.6×10^{-9} mol % detection limit would correspond to 0.02 pg/ml when referenced to the original

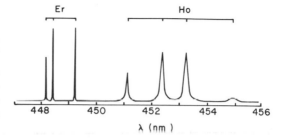

FIGURE 11. Excitation spectrum of the $^4F_{5/2}$ manifold of Er^{3+} and the $^5F_1$, $^5G_6$ manifold of Ho^{3+} in a mixed precipitate of $CaF_2:Er^{3+}$, Ho^{3+} obtained by monitoring emission from the Er^{3+} $^4S_{3/2}$ manifold and the Ho^{3+} $^5S_2$, $^5F_4$ manifold.

solution. The problem with the precipitation from more dilute solutions is that the precipitate takes longer to settle and filter, but it can be used to improve the signal levels when that is required.

The selectivity that one has in a determination is particularly important for lanthanide analyses if the separation of lanthanides from each other is to be avoided. The selectivity for specific lanthanides depends on the relative positions of the electronic levels of the different lanthanides, their line-widths, the importance of vibronics, and secondary excitation from laser-induced emissions.[9,36] In all cases the selectivity has been excellent because the lines are very sharp. Figure 11 shows the excitation spectrum of a $CaF_2:Er^{3+}$, Ho^{3+} precipitate in a region where the electronic manifolds overlap.[9] Even in this unfavorable case it is clear that the lines are well resolved. The Ho^{3+} lines are broader than those from Er^{3+} because of lifetime broadening of this particular manifold. The short lifetimes causes line broadening because of the uncertainty principle.

In the situation where two manifolds are relatively close the selectivity depends upon the vibronic contributions. The vibronic sidebands are broad and can overlap the transitions from other lanthanides. An example of the vibronic sidebands in the excitation spectrum of $CaF_2:Eu^{3+}$ is shown in Figure 12.[36] If an excitation line of another lanthanide ion falls in the region of this sideband, fluorescence can be excited from both ions simultaneously.

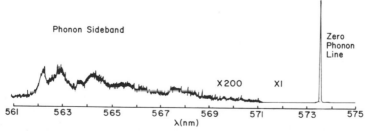

FIGURE 12. Phonon sidebands of the $^7F \rightarrow {}^5D_0$ excitation spectrum of $CaF_2:Eu^{3+}$.

A similar behavior will appear in the fluorescence spectrum. If the intensity of the vibronics is 10^3 times lower than the pure electronic transitions in both fluorescence and excitation, the selectivity will be limited to $\sim 10^6$ by the vibronics.

If the manifolds of two lanthanides are well separated and vibronics cannot contribute, the selectivity is limited by secondary excitation of the fluorescence.[9] We have observed that when a nitrogen-pumped dye laser illuminates almost any surface, a continuum of emission is generated over a wide range of wavelengths. The continuum, in turn, excites fluorescence. Typically, the fluorescence-generated off-resonance is 2×10^4 times weaker than generated on-resonance. The stray-light rejection of the monochromator determines how well fluorescence from two different ions is rejected. The overall selectivity is the product of the fluorescence selectivity and excitation selectivity. Typically it would be ca. 10^9.

The applicability of any analytical technique depends strongly on its susceptibility to interference. There are a number of different mechanisms that can produce interference:

(1) Complexation, adsorption, etc., of the lanthanide ion in solution before precipitation that changes the amount that coprecipitates.
(2) Formation of other precipitates in the original solutions.
(3) Quenching of the lanthanide fluorescence by energy transfer to the interfering ion.
(4) Formation of new sites with the lanthanide ions that compete with the site used for analysis.
(5) Changes of the site distribution because of the incorporation of other ions.
(6) Disruption of the short-range order in the lattice, causing line broadening, line shifts, and/or background.

The first two mechanisms will clearly be important in some analytical systems, but they are sufficiently well understood that we shall not be concerned with them.

A number of common ions can have strong effects on the spectra of $CaF_2 : Er^{3+}$ precipitates, as shown in Figure 13.[36] A series of $CaF_2 : Er^{3+}$ precipitates was prepared in the presence of Cl^-, K^+, and Na^+ ions and then ignited at 700°C. In Figure 13a, the Cl^- ion interferes by mechanism (6). A background has formed that probably corresponds to Er^{3+} ions distributed into a continuum of sites with random strained environments caused by the disruption of the lattice by Cl^-. In Figure 13b, the K^+ ion causes a completely new site distribution and an intense background. In Figure 13c the Na^+ ion changes the site distribution toward one characteristic of a higher temperature (compare with Figure 8d). Presumably the presence of Na^+ accelerates

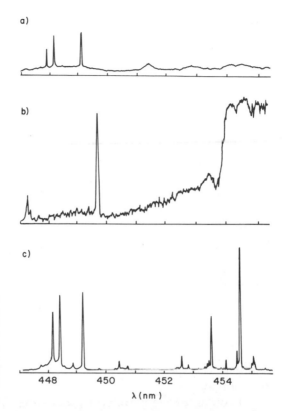

FIGURE 13. Excitation spectra of the $^4I_{15/2} \rightarrow {}^4F_{5/2}$ transition in $CaF_2 : Er^{3+}$ precipitate. If no interferences are present, only the three lines of the $G1$ site appear (see Figure 8c). The precipitates were formed in the presence of (a) $0.003\,M$ Cl$^-$, (b) $0.12\,M$ K$^+$, (c) $0.002\,M$ Na$^+$.

the conversion of fluoride to oxide compensation so the $G4$ site appears at a lower ignition temperature. There are also lines from a new site between 425 and 440 nm that come from an association of Er^{3+} and Na^+. These lines could be used for analysis of Na^+.

The interference of these ions requires that their concentration be kept low in the measurement step or that a way of preventing their effects be developed. One method of preventing other ions from changing the defect distribution is to intentionally add a high concentration of ions that define the distribution and swamp any effects other ions might have, i.e., a solid-state buffer. A number of different possible ions were tried, but the most successful method used $0.1\,M$ concentrations of $LiNO_3$ and KNO_3 in the original $Ca(NO_3)_2$ solution.[36] Precipitates formed in the presence of these high Li^+ and K^+ concentrations exhibited an acceleration of the temperature effects shown in Figure 8. The sites formed at lower ignition temperatures. The lines from the precipitates were very sharp and it appeared that Li^+ improved the ability of the lattice to perfect during the

TABLE II. Concentration of Ions for Interference in $CaF_2 : Er^{3+}$ Precipitates

Ion in solution	Concentration (M)	Ion in solution	Concentration
Ag^+	2×10^{-4}	Mg^{2+}	6×10^{-4}
Al^{3+}	10^{-4}	Mn^{2+}	4×10^{-5}
Ba^{2+}	>0.01	Na^+	4×10^{-3}
Br^-	>1	NH_4^+	>1
Cd^{2+}	2×10^{-4}	Ni^{2+}	4×10^{-4}
Ce^{3+}	4×10^{-6}	NO_3^-	>1
Ce^{4+}	2×10^{-5}	OH^-	6×10^{-3}
Cl^-	>1	Pb^{2+}	10^{-5}
Co^{2+}	>0.05	PO_4^{3-}	0.01
Cr^{3+}	2×10^{-3}	Sc^{3+}	4×10^{-6}
Cs^+	>0.1	SO_4^{2-}	2×10^{-3}
Cu^{2+}	>0.01	Sr^{2+}	2×10^{-4}
Er^{3+}	10^{-5}	Th^{4+}	6×10^{-6}
Fe^{2+}	4×10^{-5}	Ti^{4+}	4×10^{-5}
Fe^{3+}	4×10^{-6}	Y^{3+}	2×10^{-5}
K^+	>0.1	Zn^{2+}	2×10^{-4}
La^{3+}	2×10^{-5}	Zr^{4+}	2×10^{-6}
Li^+	>0.1		

ignition. The spectra became insensitive to the presence of many of the common ions.

In order to examine interferences with this solid-state buffer present, 35 ions were chosen as representative examples of potential interferences. In Table II are listed the concentration levels of these ions required for interference.[32] These values are approximate because they represent extrapolation between samples where no interference was observed and samples where it was. Ag^+, Fe^{2+}, Fe^{3+}, Mn^{2+}, and Pb^{2+} color the precipitate and lead to an intensity decrease, probably because of quenching. Ce^{3+}, Ce^{4+}, Er^{3+}, La^{3+}, Sc^{3+}, and Th^{4+} all change the site distribution by inducing the same clustering processes that caused the bending of the working curve in Figure 9. Al^{3+} and Zr^{4+} retard the kinetics of the conversion from fluoride to oxygen compensation so that the fluoride-compensated sites become more difficult to convert to oxygen-compensated ones. Cd^{2+}, Mg^{2+}, Ni^{2+}, Sr^{2+}, Ti^{4+}, and Zn^{2+} broaden the lines and cause a background fluorescence as they appear to disrupt the short-range order of the lattice.

Earlier experiments with the Li,K buffer indicated that reproducibility was a severe problem.[36] Our most recent work has eliminated this irreproducibility by lengthening the ignition time to three hours and reducing the ignition temperature from 600° to 530°C. Under these conditions, the relative standard deviation for measurement is typically 12%.

The general approach of directly exciting an ion in a sample can be extended to ions other than lanthanides. Both the actinides and many of the transition metals are known to exhibit sharp line optical transitions that are similar to those of the lanthanides.[5,12,13] Many times, the strengths of these transitions are greater than those of the lanthanides so the detection limits may be more favorable. The key to the successful development of methods for the other ions lies in finding suitable host lattices that can provide sharp lines, favorable transition moments, easy preparation, and defect distributions that are not sensitive to environmental effects. There is the additional problem that the valence state of the actinides and transition metals can change and care must be taken to assure that the proper valence state is maintained in the sample preparation.

CaF_2 is an excellent lattice for the determination of uranium. U^{6+} can be coprecipitated in CaF_2 in exact analogy to the lanthanides. After ignition, fluorescence can be observed from one site which is dominant over all others in the precipitate. The transitions from this site are strong and the detection limits are 0.4 pg/ml, a value equal to the best obtained in Table I. The effect of interfering ions is smaller for U^{6+} determinations than those listed in Table II. The alkali ions that interfere with lanthanide determinations performed without a Li,K buffer do not affect the U^{6+} determination. Thus, the Li,K buffer is not necessary in U^{6+} determinations. The interference of Al^{3+} and Mg^{2+} and other ions that retard the development of oxygen compensation in lanthanide determinations can be greatly reduced by the higher ignition temperature that is possible in a uranium determination. Ions that quench the fluorescence such as Fe^{3+} interfere at about the same level as encountered for the lanthanide determinations.

It is also possible to determine transition metals. Chromium, for example, can be determined as Cr^{3+} in $Mg(OH)_2$ and igniting the precipitate to form $MgO:Cr^{3+}$. This material has several Cr^{3+} sites but they are quite reproducible and their transitions are sharp. The detection limit for Cr^{3+} by this method is 40 pg/ml.

The extension to the actinides and transition metals (particularly the second and third row) will require considerable work because the chemistry changes more drastically from ion to ion than it does for different lanthanides. Preparation procedures will probably need to be more specific to the particular ion of interest. Nevertheless, the promise of a highly selective method with excellent detection limits is a suitable inducement for the pursuit of this goal.

2. Probe–Analyte Ion Association

The previous section dealt only with ions which were themselves fluorescent. At the beginning of this chapter, six methods were described that could cause other nonfluorescent analyte ions to associate with lanthanide probe ions and allow their analysis. The first two of those methods have not been studied. If one has a mixture of different compounds and a lanthanide ion partitions into each of them, the spectra will reflect the presence of those compounds. If the compounds were formed by various means at room temperature, we have observed that the short-range order in each is usually too low to give the sharp lines required to distinguish the different materials. If the sample is heated to allow the lattices to perfect, the ions can diffuse between compounds to create new compounds or phases. One of these phases or compounds is likely to have a much higher affinity for lanthanide ions than the others and the lanthanides will preferentially partition into it. The intensities of lines from different phases will not necessarily reflect the relative concentrations of different phases. In the general case, the result will depend critically upon what was present in the original sample. The spectra therefore become very difficult to interpret in terms of what was present in the original sample. There may very well be situations where the behavior is well defined and the lanthanide probe ions could provide valuable information on the phases that are formed or the ions that are present. There is also the interesting possibility that the natural abundance of the lanthanides in minerals could be used to determine the phases that make up a particular mineral sample. Nevertheless, this approach has not been explored experimentally.

The second method of associating analyte and probe ions was to rely upon the accidental associations that occur with a random distribution of ions. If the mole fraction of analyte ions is f_A, the mole fraction of probe ions is f_p, and if there are n positions around the probe ion that can be occupied by an analyte ion, then the probability that a probe ion will have an analyte near it will be nf_A. The mole fraction of analyte–probe pairs assuming dilute solutions then becomes $nf_A f_p$. If one can observe a 10^{-11} mole fraction (the detection limit for Eu^{3+}) of the analyte–probe pair site and one assumes $n = 6$ and $f_p = 10^{-3}$, then one would be able to observe a 2×10^{-9} mole fraction of analyte ions. This estimate would require the ability to selectively observe analyte–probe associates in the presence of a concentration of isolated probe ions 10^8 times larger. Many times this selectivity cannot be achieved and the detection limit is determined by selectivity instead. This method has the disadvantage that only a small fraction of the analyte ions are associated with the lanthanide probe ion and contribute to the signal. It is also difficult to locate the excitation and fluorescence transitions for sites that are only a small fraction of the total. The number of new sites that will appear in the spectrum will depend upon how large an influence the analyte ion exerts at the different possible lattice positions relative to the probe ion. One might expect to see multiple sites and line broadening. An example of

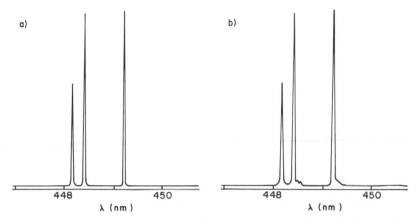

FIGURE 14. Excitation spectrum of the $^4I_{15/2} \rightarrow {}^4F_{5/2}$ transition in (a) a precipitate of $CaF_2:Er^{3+}$, (b) a precipitate of $CaF_2:Er^{3+}$, Sr^{2+}.

this method is shown in Figure 14. $CaF_2:Er^{3+}$ formed in the presence of 3 mol % Sr^{2+} exhibits line broadening and new lines close to the main intrinsic lines that are the result of association with Sr^{2+}. This method has not been pursued further because of the potential advantages of other methods.

a. Stoichiometric, Fluorescent Compounds

The probability of having an analyte ion associated with a probe ion depends upon the concentration of the probe ion, assuming a random distribution. The greatest association can be achieved by having a 100% concentration of probe ions, i.e., a pure, stoichiometric compound. Then any analyte ion that enters the lattice must enter into association with probe ions. In this limit, though, the choice of the lanthanide probe ions becomes important because energy transfer occurs readily between adjacent ions and quenching can result. Most of the lanthanide $4f^n$ levels are quenched in stoichiometric materials, but Eu^{3+} is an ion that is not usually quenched because its lower levels do not provide suitable paths for transfer.

We have examined the stoichiometric lanthanide compounds listed in Table III to determine their suitability for this application. Only half of the compounds gave sharp lines and of these only $Eu_2(SO_4)_3$ exhibited new lines when possible analytes were present. The $Eu_2(SO_4)_3$ samples were prepared by adding sulfuric acid to $EuCl_3$ solution and heating the resulting solution until the sulfuric acid fumed off. The residue was dissolved to make a stock solution that could be added to other solutions that contained the analytes. If this $Eu_2(SO_4)_3$ stock solution was evaporated to dryness at 120°C, the

TABLE III. Stoichiometric Lanthanide Compounds

Material	Sharp lines	Suitable analytes
$Eu_2(SO_4)_3$	Yes	PO_4^{3-}, AsO_4^{3-}
$EuPO_4$	Yes	—
$Eu(IO_3)_3$	No	—
$Eu_2(CO_3)_3$	No	—
$Eu(NO_3)_3$	Yes	—
$Eu_2(oxalate)_3$	No	—
$Eu_2(WO_4)_3$	Yes	—
$Eu_2(MoO_4)_3$	No	—
$Eu_2Zr_2O_7$	No	—
$Eu_2Ti_2O_7$	Somewhat	—
$EuNbO_4$	Yes	—
EuF_3	Yes	—
$EuCl_3$	Yes	—
Eu_2O_3	Somewhat	—
$EuAsO_4$	Yes	—
$Eu(BrO_3)_3$	Yes	—
$Eu_2(SeO_4)_3$	Somewhat	—
$Eu_2(TeO_4)_3$	No	—
$EuVO_4$	No	—

resulting material had the excitation spectrum for the $^7F_0 \to {}^5D_0$ transition shown on the left side of Figure 15. This spectrum was obtained by monitoring fluorescence from $^5D_0 \to {}^7F_2$ with the broad-bandpass 1/4-m monochromator. The next two excitation spectra in this figure were obtained from samples in which there was 1 mol %* PO_4^{3-} (middle trace) or AsO_4^{3-} (trace on the right) in the solution. Two new lines are found to the left of the main peak in each case. The lines in the AsO_4^{3-}-containing sample are weaker, presumably because AsO_4^{3-} enters the lattice less favorably than PO_4^{3-}. If the PO_4^{3-} concentration is lowered, the small line characteristic of PO_4^{3-} in Figure 15 rapidly disappears into the base line. If the laser is tuned to one of the excitation lines, the fluorescence spectra shown above the middle trace result. Note that the intrinsic fluorescence spectrum and the fluorescence spectrum of the PO_4^{3-}-associated site are different. It is possible to excite the PO_4^{3-} site without exciting the intrinsic site. One can now monitor the intensity of the 611.4-nm fluorescence line while exciting at 578.5 nm to analyze for PO_4^{3-}. This method has a linear working curve and a minimum detectable concentration of 0.0042 mol % or 0.3 ppm under our experimental conditions. The limitation comes from a low-level background

* Relative to SO_4 in the sample.

fluorescence from Eu^{3+} ions that are in more random environments because of lattice disorder.

If the fluorescence lines in Figure 15 are monitored and the dye laser is scanned over the excitation lines of the $^5D_1$ or $^5D_2$ manifolds (see Figure 1), the resulting excitation spectra are identical regardless of which fluorescence line is monitored. This behavior suggests that the excitation energy can be rapidly transferred between different Eu^{3+} ions when the excitation occurs above the $^5D_0$ level. The many lower ^7F_J manifolds can then participate in energy transfer. This inability to selectively excite a site to any manifold but $^5D_0$ prevents one from finding other transitions that might be stronger or better separated than $^5D_0$.

The inability to selectively excite might have an interesting application in the appropriate situation. From Figure 15 it is clear that the $^5D_0$ level in the PO_4^{3-} associated site is higher in energy than the $^5D_0$ in the intrinsic site. If the $^5D_2$ manifold of the PO_4^{3-}-associated site is excited, the energy should eventually be transferred to the $^5D_0$ level of the intrinsic site because it has the lowest energy. However, if a system had the $^5D_0$ level of the analyte-associated site lowest in energy, the excitation energy should be transferred preferentially to that level. This situation has the interesting possibility that excitation energy could be efficiently directed to the analyte transitions by

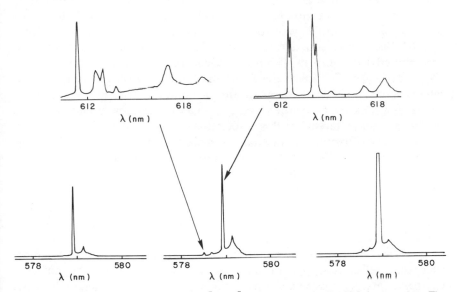

FIGURE 15. Excitation spectra of the $^7F_0 \rightarrow {}^5D_0$ transition in $Eu_2(SO_4)_3$ samples. The spectrum on the left is $Eu_2(SO_4)_3$; that in the center is $Eu_2(SO_4)_3 : PO_4^{3-}$; that on the right is $Eu_2(SO_4)_3 : AsO_4^{3-}$. The fluorescence spectra that result from exciting the two lines indicates are shown as inserts in the middle.

exciting the strong intrinsic transitions, i.e., the analyte would quench the intrinsic fluorescence and the analyte associated site would fluoresce preferentially. The analyte could itself be optically active and fluoresce directly. This type of system should have a very low detection limit but could still exhibit excellent selectivity.

An important difficulty with the idea of using pure lanthanide salts for analysis arises from the requirement that the lattice must have a high degree of perfection. Disorder in the lattice causes a continuum of randomly strained sites that result in a background which obscures fluorescence from the analyte-perturbed sites present at low concentrations. In the $Eu_2(SO_4)_3$ system this background limited the selectivity and consequently the detection limit and dynamic range for PO_4^{3-} analysis. It would appear that the requirement on lattice perfection is too stringent for the stoichiometric lanthanide materials to work efficiently as described.

b. Analyte–Probe Ion Association by Charge Compensation

It is possible to achieve a 1 : 1 correspondence between the analyte and probe ions if the two ions charge-compensate each other in an appropriate lattice. The first demonstration of this situation involved coprecipitation of Eu^{3+} and PO_4^{3-} in a $BaSO_4$ precipitate.[10] A solution containing trace quantities of PO_4^{3-} was obtained and sufficient Na_2SO_4 was added to make a 0.10 M solution. $EuCl_3$ was added to make a $5 \times 10^{-5} M$ solution. Then 0.10 M $BaCl_2$ was added to form the $BaSO_4$ precipitate. Both the co-precipitated Eu^{3+} and PO_4^{3+} require a charge compensation that can be intrinsic (Ca^{2+} or O^{2-} vacancies for example) or the two ions can compensate each other. If they do compensate each other, there will be Coulombic interaction that will favor their association and new lines should appear in the spectra characteristic of this association.

It is easiest to observe the changes in the $BaSO_4$ environments through the changes in the $^7F_0 \rightarrow {}^5D_0$ excitation spectra obtained with broadband monitoring because each line corresponds to one site. Figure 16 shows the changes that occur in the spectrum of a $BaSO_4 : Eu^{3+}$ precipitate as a function of the ignition temperature. Figure 17 shows the comparable spectrum for $BaSO_4 : Eu^{3+}, PO_4^{3-}$. One can see dramatic changes in the local environments when PO_4^{3-} is present or the temperature is changed.[10] At room temperature there are changes in the relative intensities of individual sites that reflect changes in the relative populations of the different environments, but there also appear new lines that are characteristic of the presence of PO_4^{3-}. The new lines cannot be seen in the 20°C spectrum of Figure 17 because they are in the bands around 577.2 nm, but they can be identified by selectively exciting fluorescence.[10] They are not suitable for

PO_4^{3-} analysis, though, because of their spectral overlap with other sites and poor reproducibility of relative intensity. As the preciptate is heated, the spectra change as the lattice perfects, water is driven from the lattice, and ions are able to diffuse. The bands that form about 580 nm are probably associated with lattice disorder that is created as the water and coprecipitated Na^+ leave the microcrystals. At temperatures above 800°C a new line appears at 578.2 nm that is characteristic of the presence of PO_4^{3-} and is assigned to a $Eu^{3+}-PO_4^{3-}$ association. The line has a relative intensity (when ratioed to one of the three intrinsic lines about 577 nm) that scales linearly with concentration between 0.28 and 19 $\mu g/ml$. At lower concentrations the line vanishes in a nonlinear manner. The standard deviation for this determination was 14%. At still higher temperatures the intrinsic and PO_4^{3-}-associated sites vanish and are replaced by a single, intense line that probably corresponds to a Eu^{3+} ion at a Ba^{2+} site with the charge compensation distant. At these higher temperatures one expects the analyte–probe ion association to be broken as the ions become randomly distributed in the lattice.

The spectra in Figures 16 and 17 contain transitions from Eu^{3+} ions in each of the different environments. One can obtain single-site excitation spectra by using a higher-resolution monochromator to monitor a fluorescence transition of an individual site. The $^5D_0 \rightarrow {}^7F_0$ fluorescence transition of a site must, of course, fall at the same wavelength as the $^7F_0 \rightarrow {}^5D_0$ excitation transitions shown in Figures 16 and 17. Figure 18 shows the $^7F_0 \rightarrow {}^5D_2$ excitation transitions from single sites that result from monitoring the fluorescence wavelengths indicated in the figure. Since these transitions occur between $J = 0$ and $J = 2$ manifolds, there can be a maximum of five lines if the spectra correspond to a single Eu^{3+} ion, and this is the case. Note also that the excitation spectrum of the PO_4^{3-}-associated site at 578.2 nm contains no transitions from other sites. There is a very high selectivity for the PO_4^{3-} site.

There are several serious difficulties that prevent the $BaSO_4$ system from being a competitive method for phosphate analysis at its present stage of development. The relatively high detection limit is determined by the sudden nonlinear drop in the PO_4^{3-}-associated Eu^{3+} site, presumably caused by dissociation at low concentrations and by the shifting defect equilibria that result from the absence of PO_4^{3-}. There is also a problem with achieving reproducible absolute intensities of the intrinsic lines. Broad bands appear irreproducibly in samples with irreproducible intensities. These bands are similar to those that appear in Figure 16 at temperatures above 570°C. As these bands become more intense, the narrow lines become less intense. Although the intensities of the narrow lines remain constant relative to each other (within ~14%), the wide variations of the absolute intensity cause

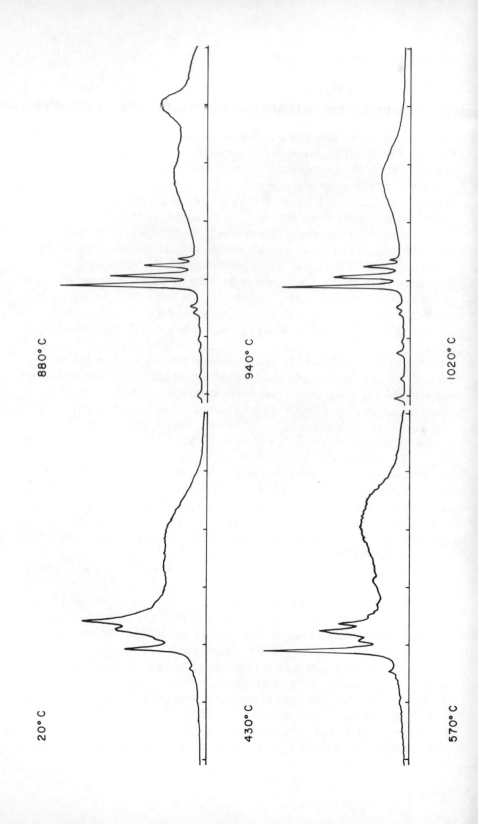

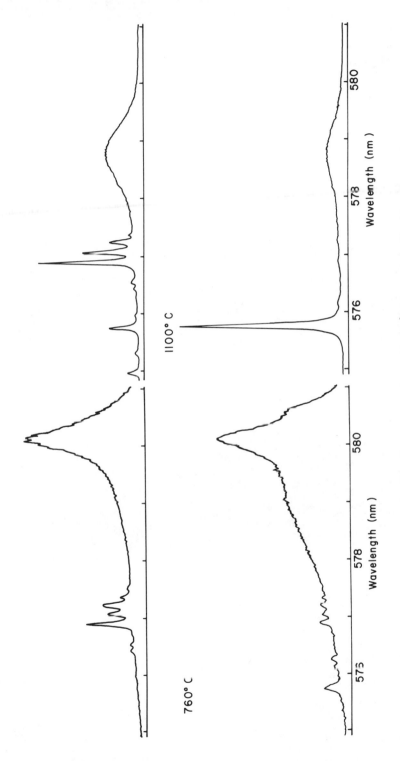

FIGURE 16. Excitation spectra of the $^7F_0 \rightarrow {}^5D_0$ transition in $BaSO_4$:Eu^{3+} for a series of ignition temperatures.

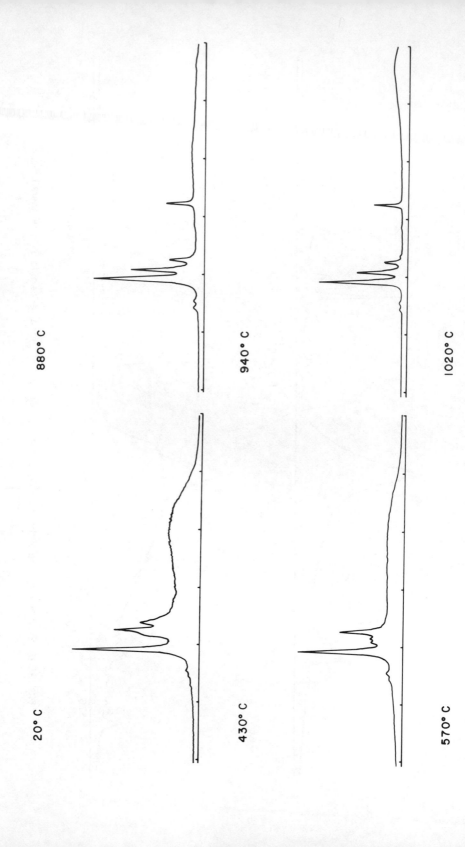

880° C

940° C

1020° C

20° C

430° C

570° C

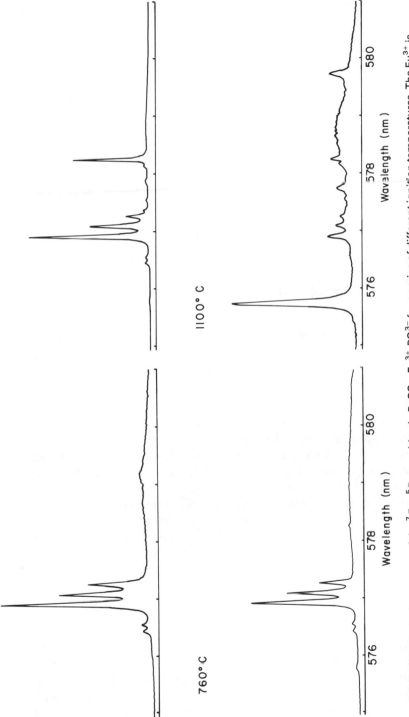

FIGURE 17. Excitation spectra of the $^7F_0 \rightarrow \,^5D_0$ transition in $BaSO_4:Eu^{3+}, PO_4^{3-}$ for a series of different ignition temperatures. The Eu^{3+} is 0.05 mol % of the Ba^{2+} and the PO_4^{3+} is 0.04 mol % of the SO_4^{2-}. The PO_4^{3-} concentration in the precipitate corresponds to a 4-μg/ml concentration in the solution.

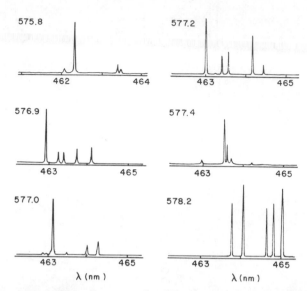

FIGURE 18. Excitation spectra of the $^7F_0 \rightarrow {}^5D_2$ transition in $BaSO_4$:Eu^{3+} and $BaSO_4$:Eu^{3+}, PO_4^{3-} precipitates that were ignited at 800°C. The fluorescence wavelength that was monitored in obtaining these spectra is indicated in the upper left of each trace. It corresponds to the transitions present in Figures 16 and 17.

uncertainties in how low a concentration can be measured for a particular sample. These variations also make it difficult to determine the effects that experimental variables exert on the sample. The detection limit, the limited dynamic range, the 14% reproducibility of relative intensity, and the irreproducible spectral changes in the $BaSO_4$ system combine to make the technique noncompetitive with other existing methods.

We have performed a number of experiments in an attempt to find conditions that would stabilize the PO_4^{3-}-associated site at low concentrations and give reproducible spectra. These experiments included trying different precipitation conditions, reagents, and methods (including homogeneous precipitation procedures), raising the samples to high temperatures to melt the $BaSO_4$, using various fluxes to improve the ability of ions to diffuse and reach their equilibrium positions, heating the samples in controlled atmospheres of O_2, N_2, He, or in vacuum, and adding other compounds to the $BaSO_4$ as solid-state buffers. Some of these procedures resulted in the spectra simplifying to a single, intense line such as appears in the 1100°C samples in Figures 16 and 17 and the disappearance of the line associated with PO_4^{3-}. These procedures gave good reproducibilities, but were not applicable for PO_4^{3-} analysis. None of the experiments were able to define procedures that would improve this method using $BaSO_4$ for PO_4^{3-} analysis.

One can next ask the question whether there are systems other than $BaSO_4$ that can have spectral lines characteristic of particular analytes but have better reproducibility and more strongly associated analyte–probe ion sites. There is unfortunately no way of answering this question without trying a large enough number of different possible systems to be representative. In general, if one has a compound $A_l B_m X_n$, where A and B are cations and X is an anion, a lanthanide probe ion and an analyte can substitute in a number of different ways. We will consider only the simplest situations. Consider first the situation where cation A is divalent and the trivalent lanthanide substitutes for A. Charge compensation can be provided by an analyte ion with one less positive charge than A or B substituting for A or B, or with one more negative charge than X substituting for X. Similarly, for the case where cation A is tetravalent, charge compensation for a trivalent lanthanide can be provided by an analyte with one more positive charge than A or B or one less negative charge than X.

A list of the systems we have studied is provided in Table IV, along with information on whether conditions could be found that gave a well-formed

TABLE IV. Charge-Compensated Materials

Material	Sharp lines	Suitable analytes	Material	Sharp lines	Suitable analytes
$BaCl_2$	No	—	MgO	No	—
BaF_2	Yes	—	NaCl	Yes	—
$BaMoO_4$	Yes	Nb^{5+}, V^{5+}, Sb^{5+}	Na_2SO_4	Yes	—
$BaNb_2O_6$	Yes	—	$Pb_5Ge_3O_7$	Yes	—
$BaTiO_3$	Yes	—	$PbMoO_4$	Yes	As^{5+}
$BaWO_4$	Yes	Nb^{5+}, V^{5+}, Ta^{5+}	$PbNb_2O_6$	Yes	—
CaF_2	Yes	Na^+, Zr^{4+}	$Pb_3Ta_4O_{13}$	No	—
$CaMoO_4$	Yes	V^{5+}, As^{5+}	$PbWO_4$	Yes	—
$CaNb_2O_6$	Yes	$Zr^{4+}, Hf^{4+}, Sc^{3+}$	$SrCl_2$	Yes	Hg^+, S^{2-}
$Ca_2Nb_2O_7$	Yes	—	SrC_2O_4	No	—
$Ca_3Nb_2O_8$	No	—	SiF_2	Yes	—
CaO	Yes	—	$SrMoO_4$	Yes	Nb^{5+}, V^{5+}, As^{5+}
$CaSO_4$	No	—	$SrNb_2O_6$	Yes	—
$CaWO_4$	Yes	Nb^{5+}, V^{5+}, Sb^{5+}	$SrSO_4$	Yes	PO_4^{3-}, AsO_4^{3-}
$CdMoO_4$	Yes	$Nb^{5+}, Bi^{3+}, Sb^{5+}, Ta^{5+}, As^{5+}$	$SrWO_4$	Yes	V^{5+}, Sb^{5+}, As^{5+}
			ThO_2	Yes	—
CdS	Yes	—	$Th_3(PO_4)_4$	Yes	—
$CdSO_4$	Yes	PO_4^{3-}	TiO_2	No	—
$CdWO_4$	Yes	V^{5+}	ZnS	Yes	Cu^+, Ag^+

lattice that had sharp lines and whether any analytes were found that produced new lines unique to the analyte ion. As can be seen, there are a number of different systems that have new lines that are associated with the presence of analyte ions. Examples of the $^7F_0 \rightarrow {}^5D_0$ transition of Eu^{3+} dopants in these materials are shown in Figures 19–24. There are four behaviors that are commonly observed when possible analytes are introduced:

(1) No changes occur in the spectra.
(2) New lines appear without changing the intrinsic spectra (e.g., Figures 23 and 24).
(3) Many new lines appear and the intensities of the intrinsic lines change dramatically (e.g., Figures 19–21).
(4) No new lines appear, but the intensities of some of the intrinsic lines decrease dramatically (e.g., Figure 22).

Despite the number of systems, none has proven suitable for a practical analysis method for reasons that are very much the same as for $BaSO_4$. Some of these systems, particularly $CdMoO_4$, ZnS, and $SrCl_2$, had spectra that were very irreproducible although sample preparation methods were kept exactly the same.

Some understanding of the problems can be obtained by looking more closely at one of these systems—$CaNb_2O_6$. Figure 24 shows the spectra of $CaNb_2O_6 : Eu^{3+}$ with and without Zr^{4+} present. These samples were prepared by sintering a stoichiometric mixture of $CaCO_3$, Nb_2O_5, Eu_2O_3, and

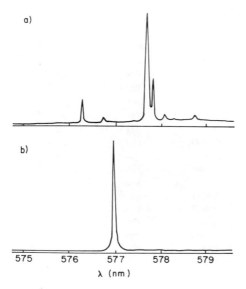

FIGURE 19. Excitation spectra of the $^7F_0 \rightarrow {}^5D_0$ transition with wide-bandwidth monitoring of a $SrSO_4$ sample containing (a) Eu^{3+} and 5 $\mu g/ml$ PO_4^{3-}, (b) Eu^{3+}.

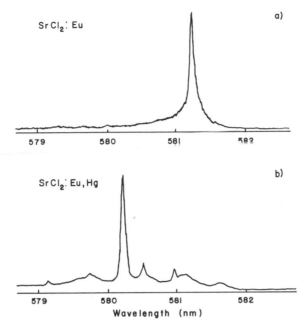

FIGURE 20. $^7F_0 \to {}^5D_0$ excitation spectra of a SrCl$_2$ sample containing (a) Eu^{3+}, (b) Eu^{3+} Hg$^+$.

ZrOCl$_2$ powders at 1100°C in Vycor crucibles a total of three times. In between firings the mixtures were ground to promote more intimate contact between reactants. The Zr^{4+}-associated line intensities are linearly proportional to concentration up to a maximum of 90 ppm Zr^{4+} (relative to the weight of CaNb$_2$O$_6$) and the Zr^{4+} detection limit was 0.6 ppm, the amount of

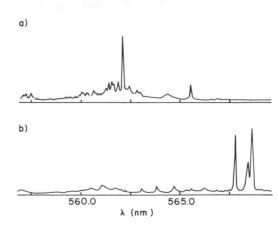

FIGURE 21. $^4S_{3/2} \to {}^4I_{13/2}$ fluorescence spectra obtained by N$_2$-laser excitation of ZnS samples containing (a) Eu^{3+}, (b) Eu^{3+}, Ag$^+$.

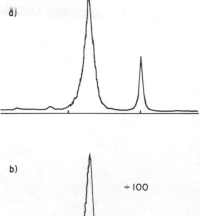

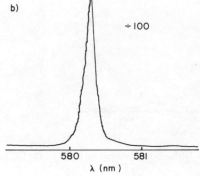

FIGURE 22. $^7F_0 \rightarrow {}^5D_0$ excitation spectra of a $PbMoO_4$ sample containing (a) Eu^{3+}, (b) Eu^{3+}, AsO_4^{3-}.

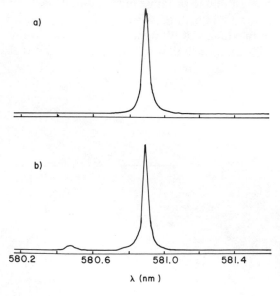

FIGURE 23. $^7F_0 \rightarrow {}^5D_0$ excitation spectra of a $CdMoO_4$ sample containing (a) Eu^{3+}, (b) Eu^{3+}, Nb^{5+}.

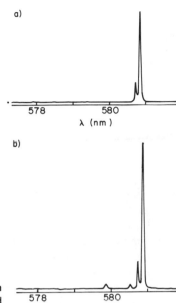

FIGURE 24. $^7F_0 \rightarrow {}^5D_0$ excitation spectra of a CaNb$_2$O$_7$ sample containing (a) Eu^{3+}, (b) Eu^{3+} and Zr^{4+}.

Zr^{4+} impurity in the Nb$_2$O$_5$ reagent. The reproducibilities of the relative intensities and the general spectral features are all less than ca. 10%.

A major drawback of this approach is the excessive sample preparation time involved in the grinding and firing steps. If these steps are reduced to a single grinding and firing, the spectra of CaNb$_2$O$_6$:Eu^{3+} have the appearance shown in Figure 25. Two separate samples are shown in this

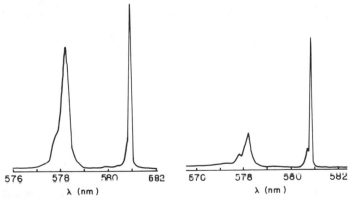

FIGURE 25. $^7F_0 \rightarrow {}^5D_0$ excitation spectra of two CaNb$_2$O$_7$:Eu^{3+} samples prepared under identical conditions, but with only one grinding and firing. Note the irreproducibility of the relative intensities and linewidths.

figure, both having the same preparation. Two new lines appear; one close to the original intrinsic line and one at shorter wavelengths, but the relative intensities and widths of these lines change irreproducibly from preparation to preparation. Clearly this behavior is not suitable for practical work. If a $CaCl_2$, Nb_2O_5 flux is used to promote more efficient diffusion, the spectrum shown in Figure 26a results. The new intrinsic line that appeared at shorter wavelengths now dominates the spectrum. If precipitation of $CaNb_2O_6$ from solution is used to form the material, the spectrum shown in Figure 26b results. All of the previous lines appear, but now many additional lines are added. These changes that occur because of different sample preparations are the result of differences in the crystalline phases and the site distributions in a phase. Only the multiple grinding and firing procedures ensure the formation of a unique phase.

If platinum crucibles are used in place of Vycor crucibles, the Zr^{4+} lines in Figure 24b decrease markedly, indicating that the crucibles themselves are part of the sample behavior. Sample contamination by crucibles is a common problem whenever one must heat materials to high temperatures. If one adds small amounts of H_3BO_3 to the samples (boron is one of the components of Vycor) in the platinum crucible, the analyte lines again appear. It thus appears that the Zr^{4+}-associated sites require B^{3+} from the

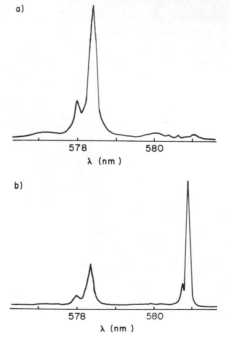

a)

578   580
λ (nm)

b)

578   580
λ (nm)

FIGURE 26. $^7F_0 \rightarrow \, ^5D_0$ excitation spectra for $CaNb_2O_6:Eu^{3+}$ samples prepared by (a) firing with a $CaCl_2$, Nb_2O_5 flux, (b) precipitation from solution.

crucibles or other sources in order to form. It is not clear whether this requirement is associated with B^{3+} catalyzing the formation of Eu^{3+}-Zr^{4+} associates because its small size promotes the mobility of ions or with actual changes in the defect equilibria caused by the B^{3+}.

It is not possible in this article to describe the detailed behavior of the remaining materials in Table IV; they all differ from each other, but they all have the sensitivity to environmental factors—crucible and furnace conditions, trace contaminants, and sample preparation conditions—in common. Thus none of these systems is suitable for practical analyses until approaches that eliminate these problems can be found.

c. Analyte–Probe Ion Association by Clustering

We saw earlier in Figures 9 and 10 that at high concentrations of Er^{3+} in CaF_2 precipitates the calibration curves departed from linearity because of the formation of Er^{3+} dimers and high-order clusters. This suggests that if other ions are present in small concentrations that have the right size and charge to enter the clusters, these ions could associate with Er^{3+} ions and cause slightly different crystal-field splittings at the Er^{3+} ions because of their slightly different size or charge. The use of clustering in this manner is the basis of the fifth case of accomplishing probe–analyte ion association described in the introduction.

The spectra for a set of CaF_2 precipitates containing 0.1 mol % Eu^{3+} are shown in Figure 27.[37] All of the precipitations were done in the presence of $0.1\ M$ concentrations of the Li^+, K^+ buffer described earlier in Section E.1. The top spectrum corresponds to a precipitate containing only Eu^{3+}, while the middle spectrum corresponds to a precipitate containing 0.1 mol % La^{3+} in addition. The line at the left (573.6 nm) is from an isolated Eu^{3+} ion with a nearby oxide compensation, while the lines at the right (around 576.4 nm) are from clusters of ions. Two additional lines appear from the Eu^{3+} ions that are in mixed Eu^{3+}-La^{3+} dimers. Their intensity is directly proportional to the La^{3+} concentration.

There are problems with this approach, because the perturbation of the La^{3+} in the Eu^{3+}-La^{3+} clusters is not sufficiently different from that of the Eu^{3+} in Eu^{3+}-Eu^{3+} clusters. The resulting differences in crystal-field splittings are not different enough to allow detection of low-concentration Eu^{3+}-La^{3+} sites in the presence of the much larger concentration of Eu^{3+}-Eu^{3+} sites. Figure 27c shows the best selectivity that can be achieved. The 1-m monochromator is tuned to a fluorescence line of the Eu^{3+}-La^{3+} site to obtain this excitation spectrum but, because this fluorescence line is partially overlapped by lines from Eu^{3+}-Eu^{3+} clusters and single Eu^{3+} sites, lines from the other sites appear in the spectrum.

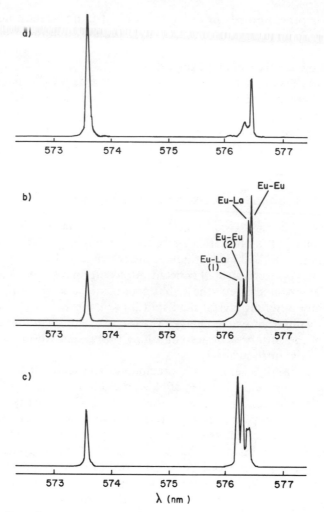

FIGURE 27. $^7F_0 \to {}^5D_0$ excitation spectra for CaF_2 precipitate samples containing (a) 0.1 mol % Eu^{3+} with broadband monitoring, (b) 0.1 mol % Eu^{3+} and 0.1 mol % La^{3+} with broadband monitoring, (c) 0.1 mol % Eu^{3+} and 0.1 mol % La^{3+} while monitoring a fluorescence line from a Eu^{3+}–La^{3+} cluster.

The solution to this problem is to select a transition from a lanthanide probe that is quenched by energy transfer when clusters containing only the probe ion form, but is not quenched when mixed clusters form.[37] Figure 28 shows the energy levels for single Er^{3+} ions (Figure 28a) and the double-ion levels for Er^{3+} dimers (Figure 28b). The double-ion level A&Y is sufficiently close to the $^4S_{3/2}$ level that efficient energy transfer occurs. In Er^{3+}–La^{3+}

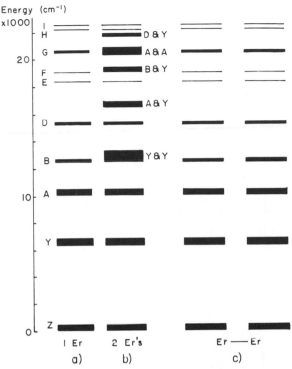

FIGURE 28. (a) Energy levels of a single Er^{3+} ion. (b) Double-ion levels of a $Er^{3+}-Er^{3+}$ dimer. (c) Energy transfer that quenches the $^4S_{3/2}$ fluorescence. The numbers beside the arrows correspond to the two energy-transfer steps that relax an excitation from the $^4S_{3/2}$ to the $^4F_{9/2}$ manifold.

dimers, however, the double-ion and single-ion levels are the same because La^{3+} has no excited levels. Consequently, the $^4S_{3/2}$ level is not quenched in the mixed dimers. The $^4F_{9/2}$ manifold (D) is not quenched to the same extent in either the $Er^{3+}-Er^{3+}$ dimers or the $Er^{3+}-La^{3+}$ dimers because the double-ion level Y&Y is further away. Consequently, the spectrum will contain only contributions from $Er^{3+}-La^{3+}$ dimers if fluorescence from the $^4S_{3/2}$ manifold is monitored, but it will contain contributions from both $Er^{3+}-Er^{3+}$ and $Er^{3+}-La^{3+}$ dimers if fluorescence from the $^4F_{9/2}$ manifold is monitored.

Figure 29 shows the expected behavior.[37] Figure 29 is the $^4I_{15/2} \rightarrow$ $^4F_{5/2}$ excitation spectrum of a 0.02 mol % $Er^{3+}:CaF_2$ sample obtained by monitoring fluorescence from the $^4F_{9/2}$ manifold. Lines from the $Er^{3+}-Er^{3+}$ dimers appear along with lines from the $G1$ site of single Er^{3+} ions compensated by one oxide ion. Figure 29b is the same excitation spectrum of

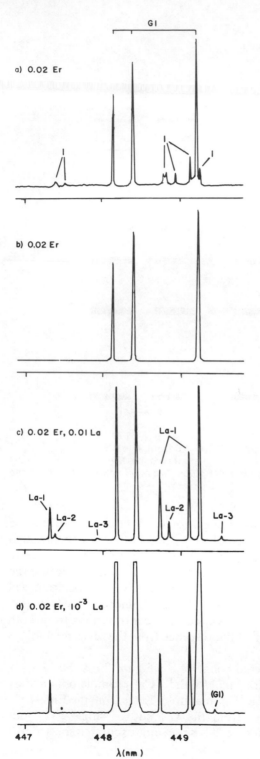

FIGURE 29. Excitation spectra for the $^4I_{15/2} \rightarrow {}^4F_{5/2}$ transition in a CaF_2 precipitate with (a) 0.02 mol % Er^{3+} obtained by monitoring fluorescence from the $^4F_{9/2}$ manifold, (b) 0.02 mol % Er^{3+} obtained by monitoring fluorescence from the $^4S_{3/2}$ manifold, (c) 0.02 mol % Er^{3+} and 0.01 mol % La^{3+} monitoring $^4S_{3/2}$ fluorescence, (d) 0.02 mol % Er^{3+} and 10^{-3} mol % La^{3+} monitoring $^4S_{3/2}$ fluorescence.

the same sample obtained by monitoring fluorescence from the $^4S_{3/2}$ manifold. The lines from $Er^{3+}-Er^{3+}$ dimers do not appear because the fluorescence is quenched from the $^4S_{3/2}$ manifold. Figures 29c and d show the same excitation spectra for samples containing different concentrations of La^{3+}. Note that only the $Er^{3+}-La^{3+}$ clusters contribute to the spectra. At higher concentrations (Figure 29c) additional lines begin to appear because $Er^{3+}-La^{3+}$ clusters of higher order than dimers form.

This procedure can be extended now to any of the other ions that can substitute in the $Er^{3+}-Er^{3+}$ clusters.[37] Figure 30b shows the same $^4I_{15/2} \rightarrow$ $^4F_{5/2}$ excitation spectrum for a $CaF_2:Er^{3+}$ precipitate that also contains La^{3+}, Ce^{3+}, Gd^{3+}, Lu^{3+}, Y^{3+}, and Th^{4+}. Figure 30a shows the $^4I_{15/2} \rightarrow {}^4F_{7/2}$ excitation spectrum for the same sample. Additional lines appear that are characteristic of each of the analytes in the sample. The lines from $Gd^{3+}-$ Er^{3+} dimers are broader than the rest because the ^8S ground state of Gd^{3+} can induce magnetic exchange splitting of the Er^{3+} states. Although these spectra look crowded, it should be remembered that they were obtained with a broad-bandpass monitoring instrument. Each of these sites can be selectively excited or monitored with the 1-m instrument in the same manner that has been described for the $BaSO_4$ and $Eu_2(SO_4)_3$ systems. This

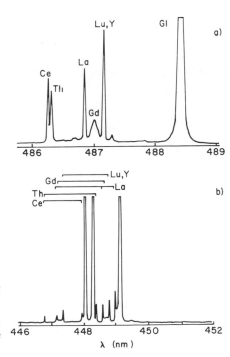

FIGURE 30. Excitation spectra of a CaF_2 precipitate containing 0.02 mol % Er^{3+} and 2×10^{-3} mol % each of La^{3+}, Ce^{3+}, Gd^{3+}, Lu^{3+}, Y^{3+}, and Th^{4+} for the (a) $^4I_{15/2} \rightarrow {}^4F_{7/2}$ transition, (b) $^4I_{15/2} \rightarrow {}^4F_{5/2}$ transition.

method therefore has the same very high selectivity that has been observed previously. Only the Lu^{3+} and Y^{3+} transitions are difficult to distinguish in the excitation spectra shown but they can be separated in the fluorescence spectrum.

The intensities of the lines in Figure 30 relative to the Er^{3+} $G1$ site are linearly proportional to the ion concentrations from 10^{-3} mol % 10^{-6} mol % of the Ca^{2+} concentration. At higher concentrations higher-order clusters begin to form (see Figure 29c) and the calibration curve becomes nonlinear. The lower limit in all cases but Th^{4+} was established by the reagent $ErCl_3$, which had a 10^{-4} impurity level of other lanthanides. The detection limits established by the signal/noise (S/N) ratio under the conditions of our experiments are listed in Table V. The levels are comparable to those obtained for direct Er^{3+} analysis because a 1 : 1 correspondence has been achieved between the Er^{3+} and the nonfluorescent analyte. The standard deviation associated with five measurements of different samples was 5%. The improved reproducibility of this method over the direct lanthanide analysis described earlier is the result of measuring intensities relative to the single Er^{3+} $G1$ site.

In order to determine the influence of possible interferences for this method, most of the ions listed in Table II were used to determine their effect on this analysis method. The results of this study are summarized in Table VI. In most of the cases the interference levels were improved by an order of magnitude and in a number of instances they were essentially eliminated. The improvement results from the measurement of relative intensities and the stabilization afforded by the higher concentration of the Er^{3+} probe ions.

TABLE V. Analytical Parameters for Nonfluorescent Ions

Ion	Excitation (nm)	Fluorescence (nm)	Fluorescence lifetime (μs)	Detection limit (pg/ml)
La	486.84	547.30	258	8[a]
Gd	487.1	547.2	253	46[a]
Lu	487.14	547.15	246	16[a]
Y	487.12	547.28	231	4[a]
Ca	486.24	545.32	388	4[a]
Th	486.30	545.83	372	8

[a] Limited to 200 pg/ml by impurity of reagent erbium.

TABLE VI. Concentration of Ions for Interference in $CaF_2 : Er^{3+}$ Precipitates

Ion in solution	Concentration (M)	Ion in solution	Concentration (M)
Ag^+	>0.1	Na^+	0.1
Al^{3+}	10^{-4}	Ni^{2+}	>0.002
Cd^{2+}	>0.01	Pb^{2+}	1×10^{-4}
Ce^{3+}	5×10^{-6}	PO_4^{3-}	4×10^{-4}
Cl^-	0.08	Sc^{3+}	6×10^{-5}
Co^{2+}	0.006	SiO_3^{2-}	0.004
Cr^{3+}	>0.02	SO_4^{2+}	10^{-4}
Cu^{2+}	0.006	Sr^{2+}	>0.02
Fe^{2+}	6×10^{-5}	Ti^{4+}	6×10^{-5}
Fe^{3+}	6×10^{-5}	VO_4^{3-}	4×10^{-4}
Mg^{2+}	1×10^{-4}	Zn^{2+}	>0.02
Mn^{2+}	6×10^{-4}	Zn^{4+}	3×10^{-5}

F. CONCLUSIONS

The research described in this chapter represents a very different approach to laser-induced fluorescence analysis and offers the advantage of improved selectivity over other methods. The use of this approach for direct lanthanide and actinide analysis and for indirect analysis of nonfluorescent ions through their association in clusters appears to be nearing the stage where it will be useful for practical analysis problems. The methods that form analyte–probe associations by other mechanisms have substantial problems with irreproducibility that must be overcome before these methods can be developed further.

This chapter has perhaps taken an overly narrow viewpoint by limiting the methods to analysis problems. The use of selective excitation of probe ions can be applied to other situations where one is interested in the short-range environment of a sample. All of the spectra displayed in this chapter contain very detailed information about the internal environments and phases of precipitates and other samples. It is clear from the majority of these spectra that many more complex phenomena occur in precipitates as a function of their formation and subsequent treatment than is evident from examining their bulk properties, such as weight or appearance, or their long-range order, such as their x-ray patterns. No concerted attempts have been made to understand the behaviors exhibited. Extensive use has already been made of the probe ion techniques to study the details of defect equilibria in model solid systems.[8,15,21,22,38,39] One should be able to apply the techniques toward studies of solution equilibria and complexation and to biological systems. It is expected that the unique information the lanthanide probe ions can provide will make their use increasingly important in present-day research where one is trying to look at the molecular level in solving and understanding important problems.

Acknowledgment

This research was supported by the National Science Foundation under Grants CHE-7424394 and CHE-7825306.

REFERENCES

1. W. M. Fairbank, W. R. Hänsch, and A. L. Schawlow, "Absolute Measurement of Very Low Sodium-Vapor Densities Using Laser Resonance Fluorescence," *J. Opt. Soc. Am.* **65**, 199–204 (1975).
2. G. S. Hurst, M. H. Nayfeh, and J. P. Young, "A Demonstration of One-Atom Detection," *Appl. Phys. Lett.* **30**, 229–231 (1977).
3. G. A. Capelle and D. G. Sutton, "Analytical Photon Catalysis: Measurement of Gas Phase Concentrations to $10^4/cm^3$," *Appl. Phys. Lett.* **30**, 407–409 (1977).
4. G. H. Dieke, *Spectra and Energy Levels of Rare Earth Ions in Crystals* (Interscience, New York, 1968).
5. W. A. Runciman, "Absorption and Fluorescence Spectra of Ions in Crystals," *Rep. Prog. Phys.* **21**, 30–57 (1958).
6. R. H. Heist and F. K. Fong, "Maxwell–Boltzmann Distribution of $M^{3+}-F^-$ Interstitial Pairs in Fluorite-Type Lattices," *Phys. Rev.* **B1**, 2970–2976 (1970).
7. R. E. Bradbury and E. Y. Wong, "Fluorescent Spectra of Sm^{2+} in KCl. I. Evidence for C_{2v} Site-Symmetry Origin of $^7F_3$ Lines," *Phys. Rev.* **B4**, 690–694 (1971).
8. D. R. Tallant and J. C. Wright, "Selective Laser Excitation of Charge Compensated Sites in $CaF_2:Er^{3+}$," *J. Chem. Phys.* **63**, 2075–2085 (1975).
9. F. J. Gustafson and J. C. Wright, "Ultra-Trace Method for Lanthanide Ion Determination by Selective Laser Excitation," *Anal. Chem.* **49**, 1680–1689 (1977).
10. J. C. Wright, "Trace Analysis of Nonfluorescent Ions by Selective Laser Excitation of Lanthanide Ions," *Anal. Chem.* **49**, 1690–1702 (1977).
11. J. C. Wright and F. J. Gustafson, "Ultratrace Inorganic Ion Determination by Laser Excited Fluorescence," *Anal. Chem.* **50**, 1147A–1152A (1978).
12. B. G. Wybourne, *Spectroscopic Properties of Rare Earths* (Interscience, New York, 1965).
13. S. Hüfner, *Optical Spectra of Transparent Rare Earth Compounds* (Academic Press, New York, 1978).
14. L. A. Riseberg and M. J. Weber, in *Progress in Optics*, Vol. 14, E. Wolf, ed. (North-Holland, Amsterdam, 1976), pp. 89–159.
15. M. P. Miller and J. C. Wright, "Multiphonon and Energy Transfer Relaxation in Charge Compensated Crystals," *J. Chem. Phys.* **71**, 324–338 (1979).
16. F. K. Fong, *Theory of Molecular Relaxation* (Wiley, New York, 1975).
17. J. C. Wright, in *Radiationless Processes in Molecules and Condensed Phases*, Vol. 15, F. K. Fong, ed. (Springer-Verlag, Berlin, 1976), pp. 239–295.

18. D. L. Dexter, "A Theory of Sensitized Luminescence in Solids," *J. Chem. Phys.* **21**, 836–850 (1953).
19. M. Inokuti and F. Hirayama, "Influence of Energy Transfer by the Exchange Mechanism on Donor Luminescence," *J. Chem. Phys.* **43**, 1978–1989 (1965).
20. N. Krasutsky and H. W. Moos, "Energy Transfer Between the Low-Lying Energy Levels of Pr^{3+} and Nd^{3+} in $LaCl_3$," *Phys. Rev.* **B8**, 1010–1019 (1973).
21. D. R. Tallant, D. S. Moore, and J. C. Wright, "Defect Equilibria in Fluorite Structure Materials," *J. Chem. Phys.* **67**, 2897–2907 (1977).
22. D. R. Tallant, M. P. Miller, and J. C. Wright, "Energy Transfer and Relaxation Phenomena in $CaF_2 : Er^{3+}$," *J. Chem. Phys.* **65**, 510–521 (1976).
23. W. B. Gandrud and H. W. Moos, "Rare Earth Infrared Lifetimes and Exciton Migration Rates in Trichloride Crystals," *J. Chem. Phys.* **49**, 2170–2182 (1968).
24. W. M. Yen, W. C. Scott, and A. L. Schawlow, "Phonon-Induced Relaxation in Excited Optical States of Trivalent Praseodymium in LaF_3," *Phys. Rev.* **136**, A271–A283 (1964).
25. R. Flach, D. S. Hamilton, P. M. Selzer, and W. M. Yen, "Laser-Induced Fluorescence-Line-Narrowing Studies of Impurity Ion Systems: $LaF_3 : Pr^{3+}$," *Phys. Rev. B* **15**, 1248–1260 (1977).
26. A. G. Walton, *The Formation and Properties of Precipitates* (Wiley-Interscience, New York, 1967).
27. F. Vaslov and G. E. Boyd, "Thermodynamics of Coprecipitation: Dilute Solid Solutions of AgBr in AgCl," *J. Am. Chem. Soc.* **74**, 4691–4695 (1952).
28. F. K. Fong, "Lattice Defects, Ionic Conductivity, and Valence Change of Rare Earth Impurities–Earth Halides," *Prog. Solid State Chem.* **3**, 135 (1966).
29. W. Van Gool, *Principles of Defect Chemistry of Crystalline Solids* (Academic Press, New York, 1966).
30. N. B. Hannay, *Treatise on Solid State Chemistry* (Plenum Press, New York, 1975).
31. C. N. R. Rao, *Solid State Chemistry* (Marcel Dekker, New York, 1974).
32. W. Hayes, *Crystals with the Fluorite Structure* (Clarendon Press, Oxford, 1974).
33. T. W. Hänsch, "Repetitively Pulsed Tunable Dye Laser for High Resolution Spectroscopy," *Appl. Opt.* **11**, 895–898 (1972).
34. J. P. Walters, *Contemporary Topics in Analytical and Clinical Chemistry*, Vol. 2, D. M. Hercules, G. M. Hieftje, L. R. Snyder, and M. A. Evenson, ed. (Plenum Press, New York, 1978), pp. 91–151.
35. C. Yang, S. Lee, and A. J. Bevelo, "Investigations of Two Trigonal (T_1 and T_2) Gd^{3+} ESR Centers in Treated Alkaline–Earth–Fluoride Crystals," *Phys. Rev. B* **12**, 4687–4694 (1975).
36. F. J. Gustafson and J. C. Wright, "Trace Analysis of Lanthanides by Laser Excitation of Precipitates," *Anal. Chem.* **51**, 1762–1774 (1979).
37. M. V. Johnston and J. C. Wright, "Trace Analysis of Nonfluorescent Ions by Associative Clustering with a Fluorescent Probe," *Anal. Chem.* **51**, 1774–1780 (1979).
38. M. D. Kurz and J. C. Wright, "Laser Excitation of Single Ion and Clustered Ion Sites in $SrF_2 : Er^{3+}$," *J. Lumin.* **15**, 169–186 (1977).
39. M. P. Miller and J. C. Wright, "Single Site Multiphonon and Energy Transfer Relaxation Phenomena in $BaF_2 : Er^{3+}$," *J. Chem. Phys.* **68**, 1548–1562 (1979).

Chapter 4

Array Detectors and Excitation–Emission Matrices in Multicomponent Analysis

Gary D. Christian, James B. Callis,
and Ernest R. Davidson

A. INTRODUCTION

Molecular fluorescence spectroscopy has received widespread acceptance as an analytical technique[1-5] The most cited advantage of fluorescence is its remarkable sensitivity,[6] especially in conjunction with high-intensity lasers as excitation sources.[7-10] Another widely cited advantage of fluorescence spectroscopy is its selectivity, especially when compared with absorption spectroscopy. However, when compared, for example, with mass spectrometry and infrared spectroscopy, room-temperature fluorescence spectra are relatively broad and featureless. Thus, the analysis of a single component or a small number of components in a complex mixture will frequently be frustrated by spectral overlaps. Well-known examples of this problem include the analysis of benzo[a]pyrene in hydrocarbon fuels and their combustion products, and the analysis of a specific drug (e.g., warfarin) in the presence of another (e.g., aspirin).

It is hardly surprising, then, that many fluorometric assays require time-consuming and tedious chemical cleanup. The frustration of the

GARY D. CHRISTIAN, JAMES B. CALLIS, and ERNEST R. DAVIDSON • Department of Chemistry, University of Washington, Seattle, Washington 98195

separation and purification step have lead to a very active effort to develop improved instrumental and curve-fitting approaches to analyze fluorescing mixtures. Selective excitation,[10] derivative spectroscopy,[11] selective modulation,[12] synchronous excitation,[13-17] low-temperature Shpol'skii effect[18,19] and site-selection[20] spectroscopy, matrix-isolation spectroscopy,[21,22] polarization measurements,[23] and nanosecond time-resolved spectroscopy[24,25] have all been used to improve multicomponent analysis. Alternatively, various numerical techniques have been used to deconvolute overlapping spectra. Foremost is the method of least squares,[26,27] which provides a useful approach to the quantitative analysis of a small number of components. Unfortunately, the least-squares approach requires that all of the components be known for which calibrated standard spectra are available.

The past few years have seen the emergence of two promising developments which can, in principle, greatly extend the capabilities of fluorescence spectroscopy as a tool for multicomponent analysis. The first of these is the availability of optoelectronic imaging detectors capable of acquiring fluorescence spectral data in large numbers of wavelength channels simultaneously.[28,29] With these devices, it is quite feasible to obtain all of the possible fluorescence emission and excitation spectra of a mixture over a wide range of wavelengths in a matter of seconds. Thus the long-sought capability of total luminescence spectroscopy[30,31] is now routinely available. The second development is in greatly improved methods for analyzing the total luminescence spectrum. This latter is most conveniently formatted as an excitation–emission matrix (EEM).[32] Under favorable circumstances it is possible for the first time to analyze a fluorescent mixture, quantitating all known components and simultaneously producing spectra of the unknown components.

It is our purpose in this review to show how these recent developments can be exploited to provide an exceptionally powerful approach to the rapid analysis of complex mixtures. Finally, we illustrate how even greater benefits can be achieved when these techniques, which we call "videofluorometry,"[33] are combined with a separation technique such as GLC, HPLC, and TLC to yield a three-dimensional approach to analysis.

B. PROPERTIES OF ARRAY DETECTORS

In this section we will review the capabilities of modern optoelectronic imaging devices (OIDs) as multichannel detectors for fluorescence spectroscopy. There has been a great deal of interest in this subject in the past five years, and there is now ample evidence that imaging detectors are no longer mere laboratory curiosities but provide significant advantages over the best available single-channel devices for almost every application of which one can think.[34]

The major appeal of the optoelectronic imaging devices is the potentially enormous multichannel advantage they offer. Succinctly stated, the signal-to-noise ratio obtained by using N detectors to simultaneously observe N channels is a factor of $N^{1/2}$ greater than that obtained by using a single detector to observe the N channels sequentially, provided that equal total observation times are employed. An alternative statement of the multichannel advantage is that the time required for N detectors simultaneously observing N channels to achieve a given signal-to-noise ratio is a factor of N less than that required for a single detector observing the N channels sequentially. The full multichannel advantage is, of course, achieved only in the ideal case where each of the N detectors is of a quality equal to the single detector, and the remainder of the system is identical. As will be shown, the present state of development of OIDs, and the inevitable optical compromises required, combine to ensure that only a fraction of the maximum multichannel advantage will be achieved.

There is another method for attaining the multichannel advantage, namely, via multiplex spectroscopy.[35] The most noteworthy achievements in this area have been in infrared spectroscopy using Michelson-type interferometers,[36] and in nuclear magnetic resonance using pulsed Fourier-transform techniques.[37] In the case of the former, interferometers also achieve a large throughput advantage over grating monochromators.[38] It can be shown that the full multiplex advantage is achieved only under circumstances where the noise is largely signal independent. Unfortunately, in fluorescence spectroscopy the noise is determined by the Poisson statistics of the photoelectrons and is thus signal-level dependent. To make matters worse, in fluorescence one is often trying to observe a very weak signal in the presence of a nearby strong signal. This situation ensures that any multiplex and throughput advantages will be negated by multiplex disadvantages.[39,40] Experimental verification of this point has been made, although there remain proponents of the approach.[41]

A second major appeal of OIDs is their capabilities in kinetic spectroscopy. This attribute of OIDs has had a large impact in more fundamental investigations and may yet prove important in analytical applications. OIDs can be used in time-resolved spectroscopic applications by (a) rapid scanning of the array,[42] (b) shuttering of the array,[43] and (c) temporally dispersing the spectral information along the array.[44] In cases where spectral information is accumulated along with temporal information, the multiplex advantage can be enormous.[45,46]

1. Photomultiplier Tubes

In order to place the various OIDs in context it is useful to review the properties of the best single-channel detector, the photomultiplier tube. One of the highest-quality general-purpose photomultiplier tubes is the RCA C31034A, shown schematically in Figure 1. Essentially, the device consists of a photocathode, which in this case is a small chip of gallium arsenide, a series of 11 copper–beryllium dynodes, and an anode, all assembled into a vacuum-tight tube with a UV transmitting window. In the usual mode of operation the photocathode is operated at a high negative voltage, each succeeding dynode is at a slightly lower negative voltage, and the anode is essentially at ground. In this device, when the photon strikes the photocathode, it liberates via the photoelectric effect a low-energy electron, which is accelerated towards the first member of the dynode chain. Each

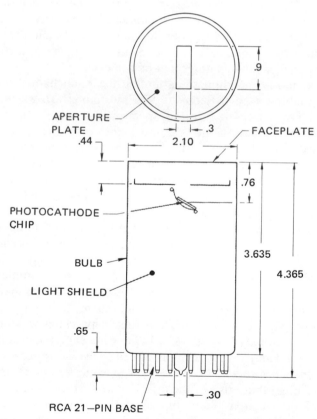

FIGURE 1. Schematic diagram of an RCA C31034 photomultiplier tube. All dimensions are in inches.

dynode, when struck with an energetic electron, releases two or more secondary electrons. These are accelerated toward the next dynode and cause emission of more secondary electrons until finally, after approximately 20 nsec, a pulse of 10^5–10^7 electrons of 2-nsec duration arrives at the anode. The resulting photocurrent is linearly proportional to the intensity of the incident light over as many as five orders of magnitude. At the high end, photocathode fatigue or anode current limitations are encountered, while at the low end, dark noises become limiting.

For applications requiring the greatest sensitivity, single-photoelectron counting techniques[47,48] are used to eliminate dark current noise arising from thermal emission from dynodes, leakage currents in the tube base, and other sources of $1/f$ noise. The remaining sources of dark noise are thermal emission of photoelectrons from the photocathode and very large pulses arising from cosmic-ray events. The latter are easily discriminated against. The former, unfortunately, give rise to pulses that are indistinguishable from photon-generated pulses. Nevertheless, it is usually possible to reduce the thermal emission rate to a manageable level by using one or more of the following strategies: (a) restricting the area of the photocathode to be as small as possible, (b) using a photocathode whose sensitivity in the red region falls off rapidly after the last red-most wavelength needed, and (c) cooling the photomultiplier tube. This last strategem must be applied carefully, as many photocathodes become less sensitive at lower temperatures, especially in the red region of the spectrum.

When the RCA C31034A PMT is operated at $-25°C$, the dark count rate is typically 25 counts/sec. The quantum efficiency of the photocathode ranges from 15 to 35% over the spectral range from 220–850 nm. If the far-red and near-IR spectral portion of the response range is not needed, one can purchase a tube with a higher quantum efficiency in the blue spectral region and achieve lower dark count rates as well.

2. The SIT Vidicon

The most widely used low-light-level multichannel device is the SIT (Silicon Intensified Target) vidicon, shown schematically in Figure 2. The heart of this detector is the 16-mm-diameter silicon target, which consists of a 1000×1000 array of islands of p-type silicon deposited of an n-type silicon wafer. Each of the islands forms a diode whose cathode is addressed by the scanning electron beam and whose anode is connected in common with all of the other diodes. In operation the target is continuously scanned, sequentially back-biasing each diode and charging its capacitance. Because of the electric field formed, majority carriers of each type of silicon are attracted away from the junction, forming a "depletion region".

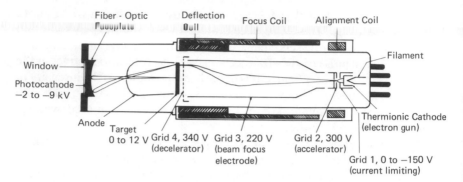

FIGURE 2. Schematic diagram of an RCA 4804 SIT vidicon tube.

When photons strike the photocathode, they cause emission of photo-electrons that are accelerated at 6–9 kV and focused by an electrostatic lens onto the silicon target. The energetic electrons penetrate to the depletion region, where they each produce approximately 1000–2000 electron–hole pairs, which then migrate to opposite ends of the diode, neutralizing the excess charges there. When the scan beam again impinges upon the target, current flows so as to recharge the diode capacitance. The amount of charge flowing through an amplifier connected to the bulk silicon will be proportional to the integral of the number of photoelectrons that fall on the particular diode between readout times.

The spectral response of the most common SIT vidicon (RCA 4804) is determined in the visible and near-IR regions by the composition of the photocathode, which is of the extended S-20 type. The ultraviolet region of the spectrum is limited by the transmission of the fiber-optic faceplate, which has a sharp cutoff at 380 nm. Response may be extended into the ultraviolet by one of three strategies: (a) A tube may be ordered with a special UV fiber-optic that extends the spectral response to ~340 nm. (b) The tube may be fitted with a relatively inexpensive proximity-focused diode-type image-intensifier tube. With this tube, one can specify both the faceplate material (quartz, MgF_2, etc.) and the photocathode (S-20, S-1, S-5, etc.). Since the standard output faceplate is a fiber optic, the intensifier is easily coupled to the input faceplate of the SIT vidicon. (c) The faceplate of the SIT tube can be overcoated with a thin layer of an organic scintillator[49] such as perylene. This highly fluorescent material serves to absorb the UV photons and convert them to fluorescence photons in the blue region of the spectrum, where the quantum efficiency of the S-20 photocathode is greatest. The only drawback to this solution is the low net quantum efficiency in the UV (~2%).

The photometric properties of the SIT vidicon when operated in the continuous scanning mode are quite acceptable.[50,51] The linearity seems to be comparable to a photomultiplier tube. The dynamic range is limited at the high end by the capacitance of the photodiodes, and at the low end by the preamplifier noise, which is about 2000 electrons per readout element. Since a single-photoelectron event can result in a charge depletion of approximately 1000, the minimum detection level is about two photoelectrons. At the high end each diode can store approximately 10^6 charge carriers, and thus a dynamic range of ~ 500 can be achieved at the highest-gain setting of the intensifier.

It is important to note that the SIT vidicon is an energy detector rather than a power detector like the phototube. Thus, it can be operated in an integration mode for improvement of the signal-to-noise ratio. In this mode the scan beam is blanked while the camera is exposed to the source of photons for a preselected time, hopefully until the signal is well above the readout noise, and then a single readout of the charge pattern is made. Thus the camera acts in much the same manner as a photographic plate. Integration time is limited by the spontaneous thermal generation of electron–hole pairs and subsequent depletion of the stored charge. At room temperature integration periods of only ~ 0.2 sec are possible; however, if the camera is cooled to dry ice temperature, exposures of up to 45 min are possible. Tremendous improvements in detection limits can be made by this technique, but it is important to note that there is a price to be paid. First, there is a small loss in sensitivity at the red edge of the spectrum. Second, the transient response of the tube is greatly reduced and exhibits much greater "lag." This phenomenon will be discussed in detail later. It is also responsible for the pronounced nonlinearity of the response of the cooled camera at low temperatures. Westphal[52] and others[53] have investigated this phenomenon in some detail. They found that for low light levels the scan beam was unable to recharge the target completely unless the target bias level is shifted by 0.6–1.0 V just before readout. While this procedure decreased the dynamic range somewhat, excellent linearity was obtained. Another important procedure is to charge the target uniformly by illuminating it to saturation and then erasing it ten times, thus achieving uniform target response. When these prescriptions are carefully followed, signal levels as low as 0.1 photoelectrons per second per channel can be observed.[54]

Thus far we have discussed the sensitivity and linearity of a single picture element (pixel). In using a multichannel detector one must inevitably deal with variations in sensitivity of the individual pixels. Much of the sensitivity variation is thought to arise from small differences in the size of each photodiode that occur at the 5% level. In addition, there is a slowly decreasing sensitivity from the center to the edges caused by the failure of the scan beam to recharge the diodes when it impinges on them at non-normal incidence. Finally, there is also a variation in dark noise from pixel to pixel. At room temperature some diodes will be totally discharged, even at scan rates of 60 Hz, giving rise to white spots in the image; often these spots can be eliminated by cooling the camera.

It seems clear, then, that to generate photometrically correct images one will have to correct for variations in sensitivity and dark current. Fortunately, for the SIT vidicon the correction is a linear one, and all one needs in addition to the observed image is an image of a uniform light source (a "flat field") and an image taken when the camera is in total darkness ("dark field"). With these images one can generate a corrected image whose pixel–pixel sensitivity variations show a coefficient of variation of less than 0.1%.[55] Once the correction algorithm is entered into the computer, the extra time needed to generate the corrected image is trivial.

The imaging quality of the SIT camera is far from perfect, but with proper insight into the source of the limitations, corrections can be made. The first difficulty arises from the "pincushion"-type distortion introduced by the electrostatic image-intensifier stage. Geometric distortion can fortunately be eliminated by a nonlinear calibration procedure using an image of a grid to generate the geometrical correction factors.[56] The resolution of the SIT camera is set by the diameter of the electron beam and the number of diodes per millimeter. The inexpensive SIT camera is capable of resolving two spectral lines of equal intensity of 50% depth if they are separated by ~50 μm. Owing to the poor quality of the electrostatic lens in the image-intensifier stage, the resolution falls off rapidly near the corners of the image. Yet another limitation of the SIT camera arises from the spread of charge carriers into adjacent pixels and the subsequent loss of resolution, a phenomenon called "blooming." Felkel and Pardue have found this effect to be particularly troublesome in the imaging of echelle spectrometers with vidicons, where it is desired to observe a weak line very close to a strong line.[57] We have not experienced any quantitative difficulties with this problem when our own SIT vidicon is employed at a resolution of 64×64 resolution elements, nor have we observed any qualitative image degradation at 256×256 resolution elements of the sort shown in Figure 2 of Ref. 57. Another potential source of cross talk is due to leakage between adjacent fiber optics in the faceplate. This effect leads to a phenomenon known as "veiling glare."[54]

The readout rate of the SIT vidicon is limited to approximately 100 frames/sec by the "lag" phenomenon. Lag is the incomplete readout of signal on a single scan arising from the exponential manner in which the capacitance is charged by the electron beam. At high light levels only 86% of the initial charge is restored by a single scan. At low light levels and at low temperatures the lag phenomenon becomes even worse. Talmi has observed that "lag" introduces a low-pass filtering operation that can be an unsuspected cause of time-response degradation.[54]

An alternative method for observing repetitive time-varying events is in conjunction with a shutter to allow a "snapshot" of an event at any particular time. Shuttering can be achieved with an electromechanical shutter or by the electro-optical technique of gating the grid of the image-intensifier section.[43] The first method is limited to times of 1 msec or longer, while the second method can achieve temporal resolutions of 40 nsec. A disadvantage of the shuttering methods is that they only allow one to obtain one spectrum per experiment. If one wants a complete set of spectra as a function of time, the experiment must be repeated at least as many times as the number of time intervals desired.

3. The Self-Scanned Diode Array

An alternative multichannel detector is the self-scanned diode array manufactured by the Reticon Corporation.[58,59] The model best suited to spectroscopic applications is the RL1024S, which comes equipped with either a quartz window or a fiber-optic faceplate, as shown in Figure 3. The device is fabricated on a single chip that contains the row of photodiodes, an array of MOS FET multiplex switches, and digital shift registers to drive the switches one at a time. A simplified schematic diagram of the device is given in Figure 4. The photodiodes are 25 μm wide, with virtually no dead space

FIGURE 3. Self-scanned photodiode array from Reticon Corp.

between them, and 2.5 mm high. This 100 : 1 aspect ratio makes them ideal for coupling to monochromators. In operation, each diode is reverse biased to 5 V in turn by connecting each MOS FET transistor to (virtual) ground through the video line. Each transistor is switched on sequentially by clocking one bit through the digital shift register. The effect of reverse-biasing the photodiode is to cause migration of the majority carriers to the electrodes, thus creating a "depletion" region at the p–n junction. In this manner the capacitance of each photodiode is charged.

For the sake of exposition, a discrete capacitor is shown in parallel with the photodiode in Figure 4. When a photon strikes the depletion region, there is a very high probability that it will create an electron–hole pair. These charge carriers then migrate in the electric field of the diode to opposite electrodes, where they neutralize the stored charge. After a chosen interval of time, the diodes are again connected in turn to the video line to reestablish the bias. The amount of charge needed to restore the bias is directly proportional to the amount of light falling on the depletion layer of each diode between bias intervals. Obviously, then, the self-scanned diode array is an integrating detector in the same sense as a vidicon.

The spectral response of the RL1024S is shown in Figure 5. Very high quantum efficiencies are obtained throughout the visible and near-IR regions of the spectrum, with a slightly smaller quantum efficiency in the ultraviolet, compared to standard photomultipliers. The falloff in the ultraviolet arises from increased attenuation by the bulk silicon, thereby decreasing the amount of light reaching the depletion layer.

The photometric properties of silicon diode arrays are excellent. Response is linear,[58,59] and for the RL1024S a dynamic range of 10,000:1 has been reported by the manufacturer. The high end of the response is limited by the finite storage capacity of the diodes, while at the low end a number of noise sources limit performance. The largest source of noise arises from capacitive coupling of clocking signals onto the common video line. This leads to the appearance of a transient on the edge of each charge

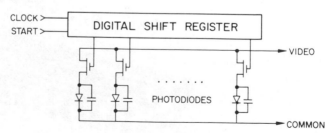

FIGURE 4. Equivalent circuit for a linear photodiode array (simplified version).

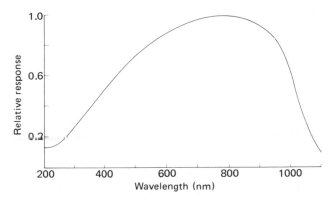

FIGURE 5. Spectral response of a linear photodiode array.

pulse. Because there are two readout clocks, and the parasitic capacitance is different from each diode, a fixed pattern offset arises. Fortunately, if careful attention is paid to producing very stable clocking waveforms, the fixed pattern noise may almost entirely be removed by subtraction of a dark frame. Another large source of noise occurs when the device is operated at room temperature: the Poisson statistics of thermal generation of electron–hole pairs in the depletion region. This source of noise may be reduced to a negligible level by cooling the device. The next source of noise arises from the preamplifier. Geary[60] has discussed a floating gate preamplifier design that apparently possesses a noise level far below that of the remaining noise sources, which arise from the chip itself. The most likely noise source is that arising from the capacitance of the video line. This noise source is probably the ultimate thermodynamic limitation to sensitivity and is greatly reduced on the RL1024S chip, compared to previous models. In practical devices it appears that the readout noise will not be reducible to much below several hundred electrons per readout.

Since the self-scanned photodiode array is an energy detector like the SIT vidicon, it can be operated in an integration mode. As with the SIT vidicon, thermal generation of electron–hole pairs limits the integration time. Vogt et al.[58] have shown that when the array is cooled to −130°C, integration times of up to 3 h can be obtained with background noise levels equivalent to one electron per second per diode. In contrast to the SIT vidicon, the tiny diode array is quite easy to cool, and special readout techniques are not required to maintain photometric linearity.

The self-scanned photodiode array possess a number of other potential advantages over the SIT vidicon. First is the reduction of lag; Vogt[58] reports that less than 1% of a signal remains upon readout a second time.

Blooming is greatly reduced as well; Horlick[59] reports data that show that blooming is virtually absent, even with large overloads. Cross talk between adjacent diodes is also very low; Livingston *et al.*[61] report that the diode arrays behave as almost ideal sampling devices in the 400–750 nm wavelength range. At longer wavelengths degradation falls off, presumably owing to the fact that lower-energy photons are absorbed deeper in the silicon substrate so the resulting charge can diffuse laterally. Finally, since the position of the diodes is fixed, there are no image distortions, and all calibrations of the diodes remain quite constant.

Except for very intense fluorescence or for near-IR measurements, the diode array will not have sufficient sensitivity for routine use, or else will require unacceptably long integration times. Two methods have been used to achieve lower detection limits. The first is to operate the array in the electron-bombarded mode, in analogy with the SIT vidicon, and the second is to fiber-optically couple an image intensifier to the array. The self-scanned Digicon[62,63] is the only commercially available version of the electron-bombarded array; it is shown in Figure 6. This device consists of a multialkali (S-20) photocathode, a magnetically focused electron-imaging section, and a self-scanned photodiode array, all assembled into a vacuum chamber. The diode array is mounted on a special aluminum header that masks off all of

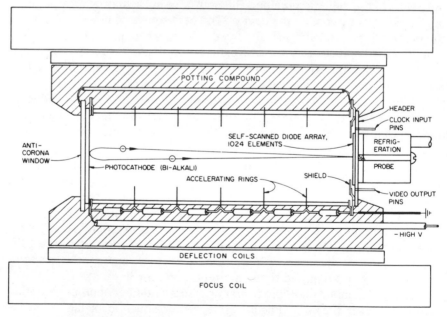

FIGURE 6. Schematic diagram of the self-scanned Digicon.

the on-chip circuit elements except the diodes themselves. The array can be cooled to $-76°C$, allowing the signal to be integrated directly on target for periods of up to several minutes. The operational principles of the Digicon are similar to those of the SIT vidicon: Photons striking the photocathode liberate photoelectrons, which are accelerated toward the array of diodes. Operation at 25–30 kV produces a photoelectron gain of 4000–5000, adequate to raise the magnitude of a single-photoelectron event well above the readout noise. The Digicon's photometric properties retain many of the advantages of the self-scanned photodiode (linear response, wide dynamic range, etc.), but provide greater sensitivity at very low light levels. Tull et al.[62] have studied the signal-to-noise ratio of this device and find true shot-limited performance top within 15% of theoretical. Mende and Chaffee[64] have studied the photon-counting capabilities of a device very similar to the Digicon and find that it is well suited to single-photoelectron counting, since the pulse heights arising from single-photoelectron pulses are well resolved from dark-noise pulses and two-photoelectron pulses.

While the Digicon closely approaches the properties of an ideal multi-channel detector in many respects, there are certain disadvantages associated with the device: (a) The cost is quite high (about $15,000), (b) there are some geometric distortions introduced by the intensification stage, and (c) the device performance gradually erodes owing to charge trapping of electrons in the diode array overcoat (1 μm of SiO_2); fortunately, this effect is reversible by annealing the array with low-energy (11 keV) electrons.

An alternative to operating the array in the electron-bombarded mode is to couple it to an image intensifier. A number of such intensified diode array devices have been constructed and described in the literature.[65–68] Shectman and Hiltner[65] have presented a pulse-counting spectroscopic detector using six stages of intensification coupled to a diode array. With this arrangement, single-photoelectron events were well above readout noise, and an elegant and inexpensive scheme was devised to locate the centroid of the pulses for photon-counting purposes. The Shectman detector obviously achieves the desired sensitivity, but suffers the disadvantages of bulk, as well as limited dynamic range, owing to large dead time losses in tthe relatively slow scan rate of the array.

McNall and Nordsiek[66] and Sandel and Broadfoot[68] have constructed low-light-level array detectors in which a single microchannel plate (MCP)[69–71] image intensifier is fiber-optically coupled to a diode array detector. Such a device is extremely compact. Moreover, it is easily assem bled, since the image intensifier comes with a fiber-optic output and the diode array can be supplied with a fiber-optic faceplate; thus one only needs to clamp the two fiber-optic faces together using index-of-refraction-matching oil or epoxy.

As Lampton[69] has discussed, there are a number of disadvantages to the first-generation image intensifiers as used by Shectman.[65] In these proximity-focused "diode" tubes each photoelectron from the photo-emissive cathode is electrostatically accelerated across a short gap, where it impinges on a phosphor screen anode. Gains of only 100× can be achieved this way, thus leading to a need for multiple stages and a consequent loss of resolution. Also, the cumbersome bulk and extremely slow recovery from overload make them inconvenient to use. The second-generation ("GEN II") image intensifiers are a marked improvement over the first generation. As shown in Figure 7, photons penetrate the fiber optic to the S-25 photocathode and are converted to electrons. The electrons are accelerated and focused with inversion onto the microchannel plate[70,71] The MCP functions to multiply the electrons by a factor of up to 10^4. The multiplied electron output from the MCP is then proximity focused onto the phosphor screen and the electrons converted back to photons. For ease of optical coupling to other devices, the phosphor is deposited on a fiber-optic plate.

The microchannel plate, 25 mm in diameter and 1 mm thick, consists of a fused bundle of capillary tubes of approximately 20 μm in diameter. The inside of each capillary tube is coated with a resistive secondary-emissive coating on its inside surface. Electron multiplication is the result of a cascade action in which the primary photoelectrons collide with the inner walls (secondary-emission coating) and cause several secondary electrons to be

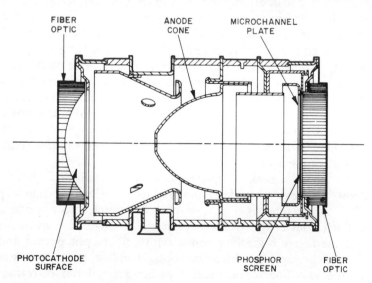

FIGURE 7. Schematic diagram of a GEN II microchannel plate image intensifier.

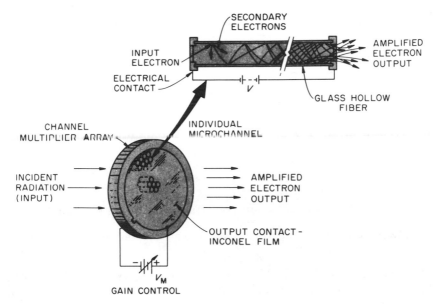

FIGURE 8. Diagram illustrating the principle of channel electron multiplication. Reprinted with permission from Y. Talmi, *Anal. Chem.* **47**, 697A (1975). Copyright by the American Chemical Society.

emitted. These in turn are accelerated down the channel, collide with the wall, and cause further emission. This process is shown schematically in Figure 8. The cascade and multiplication continues down the channel length until the end is reached or gain saturation occurs. The use of an MCP results in very high luminous gain (4×10^4–7×10^4) in a very compact package. With a current-limiting power supply, local bright areas in the image become saturated and long recovery-time "blooming" is prevented. Spatial resolution is 30 lp/mm, quite adequate for use with present diode arrays. Finally, the light output is linear with light input for light levels as low as one photon.

An alternative to the GEN II MCP intensifier is the proximity-focused MCP intensifier as shown in Figure 9. In this device the MCP is in very close proximity to the photocathode so that no electrostatic lens is needed for collecting the electrons. The advantages of this device are that (a) the input window can be constructed of a UV-transmitting substance (quartz or magnesium fluoride), thus ensuring excellent sensitivity in the UV, (b) there is no loss of resolution or sensitivity at the edges, and (c) the device is gatable in 5 nsec with on/off ratios in the 10^6–10^7 range.[72] Disadvantages of the proximity-focused MCP intensifier include slightly lower resolution, somewhat less overall gain, and somewhat shorter lifetime.

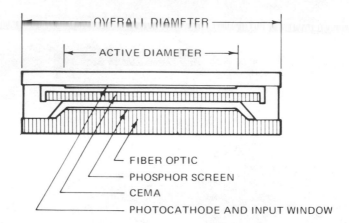

FIGURE 9. Schematic diagram of the proximity-focused MCP image intensifier.

Either of the MCP intensifiers discussed above provides enough gain to allow detection of a single-photoelectron event when a photodiode array is used as detector. However, it should be emphasized that intensified diode arrays are not at all ideal photoelectron-counter arrays. First, the problem of the event occurring in the area detected by two channels must be corrected for.[65] Second, one must realize that the pulse-height distribution of photoelectron events is a negative exponential rather than a Gaussian with a well-resolved single-photoelectron peak, as observed for well-behaved photomultiplier tubes such as the RCA C31034A. Thus, the gain process in a MCP is somewhat noisy. Schagen[73] has shown that the statistics of MCP gain follows Polya (compound Poisson) statistics. Thus, present MCP intensifiers are best operated in the analog detection mode. Photoelectron counting is in principle possible with MCP intensifiers that use the curved channel plates.[69] In these devices the microchannels are slightly curved; this allows the plate to be operated at a much higher gain without incurring ion feedback noise degradation. This latter effect is particularly troublesome when straight-through geometry channels are employed, and it arises as follows. The multiplied electrons strike residual gas molecules, ionizing them. The resulting positive ions are accelerated back to the photocathode, and collision with the latter produces a large number of secondary electrons. Curved channels prevent positive ions from reaching the photocathode and can therefore be operated at a much higher gain. In fact, the gain is so high that the pulse heights become limited by space-charge saturation effects, thus leading to a desirable Gaussian pulse-height distribution. Unfortunately, image intensifiers based upon curved microchannel plates are not yet commercially available.

As of yet, very little information is available concerning the actual performance of intensified diode arrays. McNall and Nordsiek[66] report excellent linearity of output versus integration time. Linearity of output versus input light intensity is claimed by manufacturers of intensified diode arrays, but this has not yet been independently verified. McNall further reports that the intensifier does not significantly degrade the spatial resolution of the photodiode array, that there does not appear to be any "substantial" cross talk between adjacent pixels, that there appears to be very little lag, and that one can achieve a very large intrascenic dynamic range of over 1000.

4. The Charge-Coupled Device

The charge-coupled device (CCD) imaging arrays are a promising alternative to the photodiode arrays, especially in applications requiring a two-dimensional format.[74] These devices are an array of metal-insulated semiconductor (MIS) capacitors, each serving as one picture resolution element. As shown in Figure 10, each MIS capacitor consists of a conductive electrode, an insulating layer of SiO_2, and the semiconducting substrate of p-type silicon. When a particular electrode is biased positive, a charge depletion region is formed in the bulk silicon immediately beneath the electrode. The depletion region may be thought of as a potential "well" for storage of photogenerated electrons. As for the photodiode array, the amount of charge stored is a linear function of the incident illumination intensity and the integration time. Readout is accomplished by shifting all of the packets of charge from one line of photosensing capacitors into an adjacent CCD analog shift register. Then the packets are successively shifted out to the single-readout amplifier. Shifting is accomplished by means of a three-phase clocking arrangement that varies the voltage in such a manner as to ensure the unidirectional transfer to the readout amplifier, as shown in Figure 10.

Although CCD linear and area imaging devices have been available for over seven years, there have been very few reports of the use of these devices in the scientific literature. Ratzlaff and Paul[75] have explored the use of CCD linear arrays as absorbance detectors. They found that the individual photo elements were linear in response to light level over a wide range of integration times. At high light levels the signal-to-noise ratio was over 1000/1, while at low light levels amplifier noise became limiting. Lynds[74] evaluated the photometric properties of a 244 × 199 element array detector and also found excellent linearity. With carefully designed electronics, he found that the readout noise could be reduced to a mere ten electrons! In agreement with this finding, the manufacturer of these devices has shown

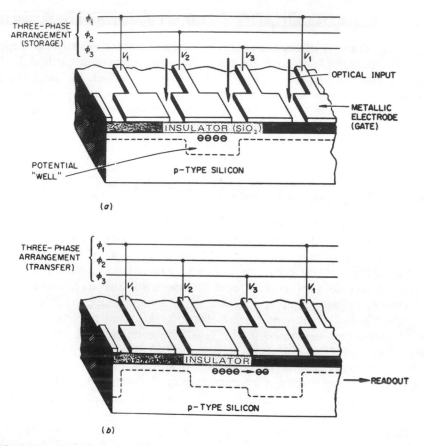

FIGURE 10. Operation of the CCD. (a) Operation in storage mode and (b) transfer mode. Reprinted with permission from Y. Talmi, *Anal. Chem.* **47**, 697A (1975). Copyright by the American Chemical Society.

imaging capabilities with only 25 electrons per photoelement.[76] The extremely noise-free properties of these chips, compared with diode arrays, arise from the fact that the capacitance of the video line is two orders of magnitude less for the CCD. Unfortunately, the commercially available CCD chips suffer from a number of disadvantages for spectroscopic imaging. First, the aspect ratio of the linear device is not favorable for interfacing to spectrometers. Second, the photons must penetrate through the gate structure to reach the active regions; this leads to a very marked loss of sensitivity in the blue region of the spectrum, with anomalous losses in certain regions of the red[74] and difficulties with "aliasing" in cases where a complex image is undersampled.

Most of the problems of the present CCDs will be alleviated when the next generation of devices being perfected by Texas Instruments become widely available. J. A. Westphal[77] has used a 500×500 photo element sensor in which the bulk silicon substrate has been thinned to 8 μm so that illumination from the backside is possible, thus greatly improving the response in the blue and ultraviolet regions of the spectrum. Westphal finds that this device, when cooled to $-40°C$ and used in the integration mode, far outperforms the cooled SIT imager he perfected some years ago.[52]

5. Comparison of Imaging Detectors and Photomultiplier Tubes— Some Theoretical Considerations

Let us begin by comparing the RCA C31034 PMT with various imaging devices, assuming that only one channel is being utilized, i.e., we will not consider the multichannel advantage, to simplify the comparison, and will assume that the PMT is operated in the photon-counting mode and that one channel of the imager is illuminated with the same number of photons as illuminate the PMT. To a sufficient approximation, the PMT signal-to-noise ratio (SNR)$_{PMT}$ can be shown to be:

$$(SNR)_{PMT} = \frac{N_p \Phi_c \Delta T}{(N_p \Phi_c \Delta T + N_d \Delta T)^{1/2}} \quad (1)$$

where N_p is the number of photons per second falling on the photocathode, Φ_c is the quantum efficiency of the photocathode, N_d is the dark count rate (apparent number of photons/sec recorded with the tube in total darkness), and ΔT is the total observation time.

In contrast, for the imaging devices presented above the signal-to-noise ratio is

$$(SNR)_I = \frac{N_p \Phi \Delta T}{[N_p \Phi \Delta T + N_p \Phi \Delta T (\sigma_g/G)^2 + N_f \Delta T (\sigma_r/G)^2 + N_d \Delta T]^{1/2}} \quad (2)$$

where σ_g is the gain noise of the intensification stage, i.e., the standard deviation of the single-photoelectron peak in the pulse-height spectrum, σ_r is the measurement noise in electrons, G is the gain of the intensification stage, i.e., the number of electron–hole pairs generated per photoelectron, N_f is the frame readout rate into the integrating digital storage device, and Φ is the ratio of photoelectrons produced to the photon input. Equation (2) can be further simplified under various assumptions. For a cooled unintensified diode array of CCD, $G - 1$, $\sigma_g = 0$, and $N_D = 0$. If the device is read out only once, then

$$(SNR)_{DA} = \frac{N_p \Phi \Delta T}{(N_p \Phi \Delta T + \sigma_r^2)^{1/2}} \qquad (3)$$

where $(SNR)_{DA}$ is the expected SNR for an unintensified diode array. For a cooled intensified diode array, $\sigma_g/G = 1$, $\sigma_r/G = 0$, and $N_D = 0$; thus, for a single readout,

$$(SNA)_{IA} = (N_p \Phi \Delta T/2)^{1/2} \qquad (4)$$

where $(SNA)_{IA}$ is the signal-to-noise ratio for an intensified diode array.

From the above equations and the characteristics of the devices as noted in Sect. B.2–B.5 it is possible to arrive at some interesting conclusions regarding the relative merits of the various devices. First, the photomultiplier tube can never be photoelectron noise limited until the rate of photoelectrons generated ($N_p \Phi_c$) becomes substantially larger than the dark count rate. Second, the unintensified diode array only becomes photoelectron noise limited when one integrates for a period long enough so that the number of collected photoelectrons greatly exceeds the square of the readout noise in equivalent numbers of photoelectrons. Third, since very long integration times are possible, signals involving far less than 1 photoelectron/sec can still be measured with photoelectron noise limitations. This is never possible with a PMT with 10 counts/sec dark noise. Under circumstances where both PMTs and diode arrays are photoelectron noise limited, the diode array is to be preferred because of its small size and its very much higher quantum efficiency, especially in the near-infrared region of the spectrum. Fourth, in order to obtain the best possible signal-to-noise ratio an unintensified diode array should only be read out once. Finally, the intensified array will be within a factor of 0.7 of photoelectron limited noise at *any* light level. Thus, one can perform multiple readouts for real-time monitoring of the signal level.

Thus far we have seen that under many circumstances the performance of multichannel image detectors is comparable to or even exceed the performance of conventional single channel detectors.[78] When it becomes desirable to scan wide spectral ranges, the time and/or signal-to-noise ratio advantages of multichannel detectors becomes enormous. However, it is important to note that a truly fair comparison must also take into account the compromises that one must make in interfacing a multichannel detector to a spectrometer. If one wants to avoid distortion of the spectral waveform, a large number of samples (at least 20) should be taken over a spectral interval equal to the bandpass of the spectrometer. Thus for a diode array with 25-μm-wide channels the slit width should be fixed at 500 μm, regardless of the ruling of the grating used. Obviously then, greater resolution (with finer-ruled gratings) is obtained only at the expense of viewing a narrower spectral range. In contrast, with a PMT as detector, one purchases the most densely ruled grating one can afford and then simply opens the slits as wide as possible, consistent with the resolution desired. Obviously, for a PMT the scanning range is limited only by the range of the grating. Thus far no critical comparison has been made of two optical systems, each optimized for the detector used.

C. APPLICATION OF ONE-DIMENSIONAL MULTICHANNEL DETECTORS IN FLUORESCENCE AND OTHER LOW-LIGHT-LEVEL SPECTROSCOPIC STUDIES

In this section we will review the use of multichannel detectors in their most straightforward configuration, that is, when used to obtain a single fluorescence or other low-light-level spectrum. Here an image of the output of a spectrograph or polychromator is focused onto the photosensitive surface of the multichannel device. For a linear array the geometry is arranged so that each of the individual elements detects a specific increment of wavelengths. If a two-dimensional detector such as an SIT vidicon is used, the signal from all elements arrayed along the dimension corresponding to the long slit axis are summed; thus one has channels whose aspect ratio matches the entrance slit.

Winefordner and co-workers were among the first to evaluate the SIT vidicon as an alternative to the PMT for steady-state fluorescence detection.[51] This group used an Aminco–Bowman SPF spectrophotofluorometer, modified to allow mounting of the SIT in place of the usual PMT, and a 150-W EIMAC-type xenon arc lamp as an excitation source. With the gratings available, a 60-nm spectral region of the fluorescence was imaged onto 500 channels. The image detector was a model 1205 OMA from Princeton Applied Research operated at room temperature. With this system a number of instructive results were obtained: (a) by summing N scans together in the digital memory of the device, the signal-to-noise ratio could be improved by a factor of $N^{1/2}$; (b) the spectral resolution for narrow spectral features (those with half-widths covering only a few channels) was within a factor of 2 from the limitation set by the entrance slit; (c) the SIT detector could cover an intensity dynamic range of 3–4 orders of magnitude without changes of gain settings; and (d) limits of detection for selected polyaromatic molecules were in the sub-ppb range. In a later paper, Cooney *et al.*[79] compared detection limits using the SIT as a parallel detector and a 1P21 photomultiplier as a single-channel detector with sequential linear scan. For equal data acquisition times of ~30 sec for equal spectral ranges these authors obtained a detection limit for anthracene approximately five times lower for the phototube. This result seems very disappointing and would not have been predicted on the basis of considerations presented in Sect. B.2 and B.5, wherein it was shown that on a per channel basis the SIT and PMT would be expected to achieve nearly equal signal-to-noise ratios for equal acquisition times. Thus, because of the multichannel advantage, one might have expected lower detection limits for the SIT camera. The comparison does not seem quite so disappointing if one takes into account the fact that the 500-channel SIT greatly oversampled the spectrum, and that summation of at least ten adjacent channels would be justified. According to Talmi *et al.*,[54] this would be expected to improve the SIT detection limit by at least a factor of 3. Yet another factor apparently not taken into account by Cooney *et al.* was the sensitivity variation from channel to channel that leads to a fixed-pattern noise not removed by subtraction. This systematic noise is easily mistaken for random noise and makes itself most apparent when first- and second-derivative spectra are displayed.[54]

An excellent and critical account of the use of an SIT vidicon in molecular fluorescence spectroscopy has been presented by Talmi.[54] Two studies contained in this paper are of particular interest. The first illustrates the signal-to-noise ratio improvement to be obtained by cooling the camera and integrating the signal directly on target. A quite acceptable spectrum of a neon lamp is shown in which the maximum intensity is only 0.1 photoelectrons per second per channel. Integration for 2000 sec was made possible by cooling the camera to $-50°C$. Unfortunately, all fluorescence spectra shown in this paper were recorded with the detector operating at room temperature; under this condition the SIT vidicon appears to offer no signal-to-noise ratio advantage over a sequentially scanned PMT detector. Mathies[80] has shown that if one is willing to bear the inconvenience of low-temperature operation, the cooled SIT vidicon can achieve a large fraction of the theoretical multichannel advantage over a sequentially scanned PMT detector. For Raman spectra, a time advantage of a factor of ~30 was obtained.

The second study in Talmi's paper was designed to test the use of a multichannel detector for simultaneous multicomponent analysis. A three-component mixture of anthracene, diphenylstilbene, and pyrene was studied in which the concentration of the last was 20 times less than that of both the others. Even though the spectral overlap of the three components was severe, good quantitation to ~5% was achieved. Such results are very encouraging, and one hopes soon to see more reliance placed on fast computer-based methods to achieve multicomponent qualitation and less reliance on time-consuming and tedious chemical cleanup.

Parallel detectors are ideal devices for transient spectroscopic studies. Of major interest to analytical chemists is the possibility of characterizing the effluents of a liquid chromatograph "on the fly."* Here the analogy to the GC–MS and GC–IR is obvious, and somewhat similar benefits might be expected. Jadamec et al.[81] have evaluated the use of an SIT detector to record the spectrum of petroleum oil aromatic hydrocarbons separated by a high-pressure liquid chromatography system. The commercially available detection system with floppy disk mass storage was capable of recording one spectrum every 10 sec.

Of critical importance was the optical flow cell.[82] In early work a 10-μl cylindrical cell was found to severely degrade the signal level and greatly increase the scattered light when compared to a standard 10×10-mm cuvette. In contrast, a specially designed square cuvette of 20-μl volume actually gave larger signals than the standard cuvette. With this system, 4 ng of chrysene produced a very acceptable signal-to-noise ratio. Moreover, a careful comparison of spectra taken of ovalene and p-terphenyl showed them to be virtually identical to spectra taken with a conventional instrument with an identical resolving power.

The most interesting of the studies reported by Jadamec were those related to the characterization of petroleum samples for legal purposes. In this work, very simple and rapid cleanup procedures were employed, resulting in correspondingly complex unresolved chromatograms. Although the fluorescence detector greatly undersampled the data, it was nevertheless possible to identify a number of peaks in the chromatogram. Especially noteworthy was the simultaneous use of retention times and spectral data to establish that a particular scan was 1-methylfluorene and not fluorene.

An excellent account of the use of the SIT vidicon for time-resolved phosphorescence studies has been given by Goeringer and Pardue.[83] Their instrument consisted of a xenon flash lamp to irradiate the sample, a flat-field spectrograph to disperse the resulting radiation, and a computer-controlled SIT vidicon, capable of recording a 300-nm spectrum in 125 channels in a time range selectable between 8 and 65 msec in 8-msec steps. An electronic shutter in front of the spectrograph prevented excitation light from reaching the vidicon during the flash excitation. Before embarking on their phosphorescence studies, Goering and Pardue measured the dynamic response of the SIT vidicon by applying a step-function light change to the system. A plot of the time (or number of scans) required to recharge the vidicon to 95% of its original level was found to depend upon the inverse

---

* This subject is also considered by Froehlich and Wehry in Chapter 2 of Volume 3 of this series.

square root of the signal size. At very low signal levels 500 msec was needed
to recharge the signal to 95% of its original level. For the compounds and
concentrations used in the present study the response time of the camera was
negligible; it cannot be overemphasized that measurement of phosphorescence decay times of 50 msec and shorter, over wide dynamic ranges,
will require very complex corrections.

Nevertheless, the present study can be considered to be quite successful
in demonstrating the potential advantages of a multichannel detector in
mixture analysis. The first advantage is the convenience of using an internal
standard to reduce the effects of experimental variables, in this case by
factors of 2 to 5. The second advantage is the use of all data points
simultaneously to obtain the best (in the least-squares sense) values of initial
intensities (concentrations) and decay times.

One very appealing use of a multichannel detector has been in a laser
radar system capable of remotely detecting oil spills in sea water.[84] The
system consists of a pulsed Nd:YAG laser as an excitation source, a
telescope for collecting the backscattered radiation, and a polychromator
and an SIT vidicon as a multichannel detector. In order to operate the
system during the daytime the image-intensifier stage is gated on only during
the laser flash. This scheme reduces background noise due to sunlight
scattered from the water surface to a negligible level. When the second
harmonic of the laser is used ($\lambda = 532$ nm), both Raman (CH stretching)
and fluorescence are observed. When the fourth harmonic was used
($\lambda = 266$ nm), only fluorescence could be observed. When field trials were
conducted, it was observed that sea water itself contributed a pronounced
peak at approximately 690 nm. This signal was undoubtedly due to the
fluorescence of chlorophyll contained in the algae. Clearly such a system can
be of enormous benefit in remote sensing of the environment and in remote
sensing for industrial process control.

Multichannel detectors have been found to be quite useful in fluorescence microscopy. Such instruments have been described by Hirschberg et
al.[85] and Jotz et al.[86]* The optical scheme of both these systems is similar.
The excitation source is an arc lamp with glass and interference filters to
isolate the desired band of wavelengths. A dichroic beam splitter is placed at
an angle of 45° to the optical axis of the microscope and serves the dual
function of (a) reflecting the exciting light through the objective, focusing it
onto the sample and (b) transmitting fluorescence radiation collected by the
objective onto the entrance slit of the analyzing polychromator. In both
cases an SIT vidicon was used to collect the entire spectrum of emitted light

---

* Kohen and co-workers discuss their elegant work in this area in greater detail in Chapter 7 of
  Volume 3 of this series.

at one time. The system of Hirschberg *et al.* also has the capability of measuring the spatial distribution of the fluorescence along a narrow striplike portion of the cell. These authors have used their system to study the uptake and metabolism of benzo[*a*]pyrene by living cells, as well as the alteration of metabolic functioning of the cell after microinjection of substrates. Jotz *et al.* have used their system to study the specific and nonspecific binding of DNA fluorochromes in mammalian cells. These authors found multichannel detection to be especially advantageous in the case where rapid fading of the fluorochromes occurred.

A novel extension of a microfluorescence spectrometer has been reported by Wade *et al.*[87] In their instrument, cells stained with a fluorochrome are caused to flow one at a time past a focused laser beam, and the fluorescence is collected at right angles and dispersed by a polychromator onto the photosensitive surface of an SIT or ISIT vidicon. With the latter detector the intensifier could be gated on when a cell passed into the laser beam, as detected by scattered or fluorescent light sensed by a photomultiplier tube. Unfortunately, it was not possible to record the fluorescence spectrum of a single cell because of the rapid flow rates employed. However, it was possible to obtain the average spectrum of several thousand cells in the course of 1 min. It would be very interesting to apply this technique to the analysis of air particulates to correlate fluorescence and Raman spectra with particle size on a particle-by-particle basis.

Thus far only one report[88] has appeared on the use of intensified diode array detectors in analytical fluorometry and chemiluminescence. Ryan *et al.* used a commercially available 512-element Reticon array fiber-optically coupled to a second-generation image intensifier whose fiber-optic faceplate had been coated with an ultraviolet scintillator to enhance sensitivity. With a small $f/3.5$ polychromator containing a coarsely ruled grating, a spectral range of 600 nm was covered, sampling at 1.5-nm intervals. With this system and an arc-lamp excitation source, a fluorescence spectrum of 0.2 ppb quinine could be recorded with a signal-to-noise ratio of 2 at the peak of emission. In contrast, a mechanically scanned monochromator with a PMT detector required a 13-min scan to obtain a spectrum of equal signal-to-noise ratio. This result is in excellent agreement with considerations based upon the ability of the diode array to detect a single-photoelectron event as outlined in sect. B.5. Studies of various chemiluminescent reactions confirmed the capability of the array detector to obtain spectral data on transient species with good sensitivity.*

---

* The use of diode arrays for the performance of kinetic-based analyses by fluorescence and chemiluminescence is considered in greater detail by Ingle and Ryan in Chapter 3 of Volume 3 of this series.

Our own experience[89] with this detector–polychromator has also been quite gratifying. Using an argon-ion laser as an excitation source, we were able to obtain a detection limit for Rhodamine 6G of less than 1000 molecules. However, calculations indicated a detection limit of one molecule, and indeed the detector was operating on its lowest gain range, indicating a capability of achieving two orders of magnitude better detection limits. The reason for the inability to achieve theoretical limits became quite apparent when a high-resolution monochromator was substituted for the low-resolution instrument. Minute particles of dust were randomly diffusing into the laser beam and emitting large emounts of Raman and fluorescence light. The high variability of Raman and fluorescence spectra from such particles produced a variable background spectrum that could not be subtracted out. Future studies are aimed at removing particulates from the solution and thereby greatly improving the detection limits.

D. THE EXCITATION–EMISSION MATRIX

In this section we present an unconventional way of thinking about fluorescence analysis that is particularly fruitful in dealing with mixtures.[90] Fluorescence intensity is, of course, a function of both the wavelength of excitation (λ_{ex}) and the wavelength of emission (λ_{em}). Generally, one wavelength is held constant and the fluorescence intensity is recorded while the other is scanned; thus either an excitation or emission spectrum may be obtained. Because fluorescence at room temperature generally occurs from the lowest vibrational level of the first excited singlet state (owing to very fast radiationless decay to this level), the relative shape of the emission spectrum is independent of the wavelength of excitation. Only its intensity will vary. Conversely, since the fluorescence quantum yield is generally independent of the wavelength of excitation, it follows that the excitation spectrum is independent of the monitored emission wavelength. Obviously, then, to describe the fluorescence of a dilute solution containing a single species one would need to scan only two spectra; one emission spectrum and one excitation spectrum. However, if a solution containing two independently emitting species is examined, the emission spectrum is found to be *dependent* upon the wavelength of excitation; also, the excitation spectrum is now dependent upon the wavelength of emission monitored. In the absence of independent knowledge of the specta of the fluorescent species, it would be necessary to acquire the fluorescence emission data in a systematic fashion as a function of a sequence of emission and excitation wavelengths. Weber was perhaps the first to realize this and proposed to represent the data in matrix form, as an excitation–emission matrix (EEM), in which each matrix element is the fluorescence intensity at particular values of the wavelength of excitation and emission.[91]

1. Construction of the EEM

In order to better understand the nature of the EEM, let us consider a mixture of two fluorophores with spectra as shown in Figures 11A and B. In Figures 11C and 11D are the EEMs for each component shown graphically in the form of a contour plot, with each contour line representing a particular value of fluorescence intensity. The diagonal lines across the EEMs represent the scattered radiation from the excitation source that will be most intense when $\lambda_{ex} = \lambda_{em}$. All the fluorescence is constrained to lie on or below this line (Stokes' law). A horizontal slice across the EEM represents the emission spectrum at the particular wavelength of excitation, and a vertical slice represents the excitation spectrum at the particular wavelength of observed emission. Hence, because the emission spectrum is independent of the excitation wavelength, we would expect to observe for component 2 two emission peaks (one strong and one weak) at each of the wavelengths of excitation (each horizontal slice), the spectra being more intense at the wavelengths corresponding to the excitation peaks. The excitation peaks in this case are more intense at the shorter wavelengths. Conversely, we would expect four excitation peaks (strongest at the blue end) at each wavelength of emission (vertical slice), being most intense for observation at the 625-nm peak.

In Figure 11F is shown an EEM for a 1:1 mixture of the two fluorescing components; now the emission and excitation spectra are no longer independent of the wavelength of excitation or observation. Provided that there is no appreciable electronic energy transfer between species and that they emit independently (dilute solution), the EEM of Figure 11F is simply the sum of C and D. In Figure 11E we give an alternative representation of the EEM—an isometric projection with all hidden lines removed for clarity.

2. Characteristics of the EEM

Various conventional and less conventional fluorometric techniques can be visualized by means of the EEM.

a. Selective Excitation and Emission

An important feature of the EEM is the unique spectral regions where each component absorbs and emits independently of the other. Most routine analyses are based on this feature, i.e., selective excitation and emission

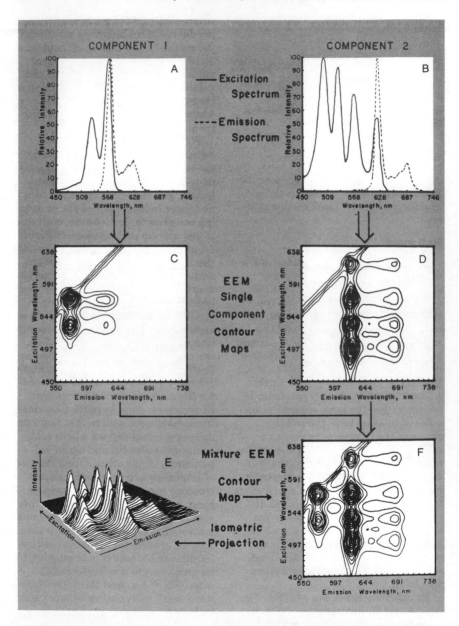

FIGURE 11. Illustration of relationship of EEM to a compound or mixture's excitation and emission spectra. Reprinted with permission from D. W. Johnson, J. B. Callis, and G. D. Christian, *Anal. Chem.* **49**, 747A (1977). Copyright by the American Chemical Society.

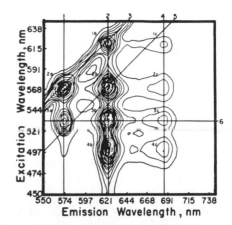

FIGURE 12. Relationship between EEM and conventional fluorescence spectral scans.

techniques. These techniques rely on the analyst's ability to locate regions in the EEM in which the component of interest is spectrally isolated from other emitters and has sufficient signal for a quantitative measurement. That is, a particular wavelength of excitation and of emission is selected at which only the compound of interest fluoresces. Other components of a mixture may be determined by selecting other appropriate wavelength pairs. Figure 12 demonstrates that peak 2a would be the best choice to measure component 1 for the mixture in Figure 11. That is, the fluorescence would be monitored at 574 nm with excitation at 568 nm. The excitation spectrum could be recorded by monitoring emission at 574 nm (vertical slice, line 1). But an emission scan (horizontal line) would result in an overlap with the spectrum of component 2. The quantitative measurement of component 2 would be difficult, since only its weak emission peak at the longer wavelength can be measured (on line 4). Peak 4c would be the best.

Any point or line on the EEM provides only these particular pieces of information, leaving the analyst "blind" to other regions of the EEM that may be more suitable for the sought-for component or that may contain information as to the nature or signal level of an unexpected interfering compound.

b. Synchronous Excitation

Lloyd[13,14] introduced the synchronous excitation technique for qualitative and quantitative analysis of mixtures.* The excitation and emission monochromator drives are locked together at a fixed wavelength interval

---

* The use of synchronous luminescence spectrometry in analytical chemistry is discussed in detail by Vo-Dinh in Chapter 5 of the present volume.

($\Delta\lambda$), e.g., 20 nm, so $\lambda_{em} = \lambda_{ex} + \Delta\lambda$. The synchronous-scan concept takes advantage of the variability of both the excitation and emission spectra of fluorescent species simultaneously, i.e., it explores different areas of the EEM. The synchronous-scan spectrum can be easily visualized on the EEM as a line parallel to, and to the red of, the scattered excitation radiation, where $\Delta\lambda$ is the interval of offset (e.g., line 5 in Figure 12). The resulting spectrum contains only a few peaks that are slices through the "0–0" transitions of the components of the mixture (the transitions for which the wavelength of emission is equal to the wavelength of excitation). The factors influencing the shape and locations of these peaks are described in detail in the papers by Lloyd,[15] John and Soutar,[16] and Vo-Dinh.[17]

The technique has found use in fingerprinting polyaromatic hydrocarbon mixtures and crude oils.* It works well for mixtures whose components possess well-defined vibronic bands. A small $\Delta\lambda$ (3 nm) is needed to produce simple spectra for structured components in a mixture.[17] The technique is less specific in cases of unstructured spectra or where the 0–0 transition is less probable.

The synchronous-scan method has the same blindness property for locating optimal regions of the EEM, as mentioned previously, and may miss certain components in a mixture.[90]

E. ACQUISITION OF THE EEM

Various approaches have been taken to acquire excitation–emission matrices. Weber,[91] in his pioneering description of the EEM, simply took a few selected emission readings manually at a few selected excitation settings to obtain a low-resolution 3×3 or 4×4 EEM. With the advent of the inexpensive laboratory computer, a number of investigators have described automated instruments well suited to obtaining an EEM.[92] In one of these instruments the computer controls the scanning of the monochromators with stepping motors, and fluorescence intensities are acquired through digitization of the output of the analog photomultiplier signal. Scanning is accomplished by holding the excitation wavelength constant while acquiring a 50-point emission spectrum, then changing the excitation wavelength, and recording another fluorescence spectrum until 2500 points are acquired. The data are formatted and printed out in the form of an EEM, either as a contour plot, as in Figure 1, or a three-dimensional isometric projection.[92]

---

* Various luminescence approaches (including synchronous excitation) for identification of oil samples are considered by Eastwood in Chapter 7 of the present volume.

Hornig and co-workers[93] also developed an EEM acquiring instrument that is now commercially available. This computer-controlled instrument, with an IBM contouring package, allows different manipulations of the stored data.[94] Difference spectra can be obtained, for example, to subtract solvent background and scattered light. The intensity grid is normalized to 1000, and a minimum (usually 50) and maximum (usually 1000) contour value and a contour interval are selected. A contour interval of 50 would give 20 contours. The minimum and maximum contour values can be changed, for example, to chop off the top of a strongly emitting component that may be obscuring a minor component.

Other suitable instruments have been described by Haugen et al.,[95] O'Haver et al.,[96] and Rho and Stuart.[97]

F. THE VIDEOFLUOROMETER

Clearly, it would be highly desirable to have the capability to routinely obtain an EEM on every sample.[93] Unfortunately, with conventional instruments the long time needed to acquire this data precludes routine use. In most cases rapid mechanical scanning is not feasible because of reproducibility problems and loss of signal-to-noise ratio due to the decreased number of photons counted.

These difficulties with using conventional instrumentation to obtain the EEM prompted us to explore alternative methods for data acquisition. Thus in 1975 Warner et al.[33] described a way to obtain the entire EEM simultaneously using an array detector. The system is based on illumination of the sample with polychromatic radiation rather than monochromatic radiation.

1. Principle of Simultaneous Acquisition of EEM

Consider a hypothetical compound with excitation and emission spectra as depicted in Figure 13A. Suppose a cuvette containing a solution of this compound is illuminated by polychromatic radiation as in Figure 13B. One would expect three bands of fluorescence corresponding to the wavelengths of the three excitation peaks. The emission bands would all be the same color, since they all arise from the same excited state of the same molecule. The top band would be the most intense, corresponding to excitation at the wavelength of the green excitation peak (the largest one at 500 nm). Excitation at the small violet peak would result in the weakest emission band (bottom of cuvette).

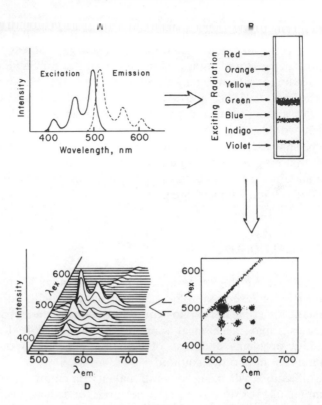

FIGURE 13. Simultaneous production of EEM for a hypothetical compound. (A) Excitation and emission spectra. (B) Fluorescent cuvette as illuminated by polychromatic beam. (C) EEM as viewed by SIT camera. (D) Isometric projection of EEM as may be obtained on an oscilloscope or graphics terminal. Reproduced with permission from D. W. Johnson, J. A. Gladden, J. B. Callis, and G. D. Christian, *Rev. Sci. Instrum.* **50**, 118 (1979).

Now suppose that each band is dispersed horizontally by means of a diffraction grating or prism to give the emission spectrum. The result would be three emission peaks at each band, the shorter-wavelength peak being the most intense in each case. We would then have a representation of the excitation–emission matrix, as in Figure 13C. In this case there is a 3 × 3-peak EEM, resulting in nine peaks. Note also the scatter line. The most intense peak is the 0–0 transition, since the most intense excitation peak and most intense emission peak both occur here.

The videofluorometer presents such an EEM simultaneously to an array detector.

2. The Polychromator–Detector System

The instrumental setup for the videofluorometer is shown in Figure 14. A 150-W xenon arc lamp is used as a continuum source that is dispersed in the excitation monochromator. The exit slit and its holder are removed, however, so that polychromatic radiation is focused on the cuvette. The instrument is placed on its side or end, depending on the design, so the dispersed radiation (260-nm span) falls vertically on the cuvette. This produces the bands of fluorescence, as was illustrated in Figure 13, that are then dispersed horizontally in the analyzing monochromator. The long axis of the slit of the analyzing monochromator is oriented parallel to the cuvette in the usual fashion. Since the vertical image at the entrance slit is maintained through the emission monochromator, the excitation information is preserved. The emission monochromator is also converted into a polychromator by removing the exit slit and holder. Since the fluorescence present along each vertical position of the entrance slit is

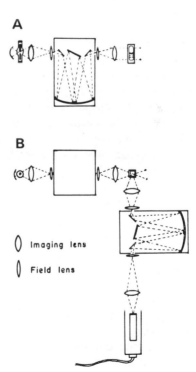

FIGURE 14. Optical diagram of the video fluorometer. (A) Side view of excitation beam. (B) Top view detailing emission path to camera. Reproduced with permission from D. W. Johnson, J. A. Gladden, J. B. Callis, and G. D. Christian, *Rev. Sci. Instrum.* **50**, 118 (1979).

horizontally dispersed into its component wavelengths then the two-dimensional EEM is presented at the exit plane.

An excellent account of the optical considerations that must go into the design of an imaging fluorometer is given by Warner.[98]

In the University of Washington videofluorometer conventional high numerical aperture Czerny–Turner monochromators were employed. These monochromators were never designed for imaging purposes and certain distortions of the EEM result. The most severe problem arises from the considerable astigmatism that is a natural result of high numerical aperture being traded off for resolution along the slit length. We found that placing an aperture stop at the focusing lens in front of each monochromator so that only one half of the available area of the collimating mirrors is used decreased the aberrations to the point that acceptable resolution (~5 nm) was obtained at all points on the image plane. Additional minor sources of degradation in our system arise from chromatic aberration due to the use of single-element lenses as focusing elements, coma, and reentrant spectra. A final drawback to the present system is the fall-off in sensitivity arising from the inability to capture all of the off-axis rays; this phenomenon is often referred to as vignetting.

Warner has presented an improved optical design for a videofluorometer.[98] In his scheme commercially available concave holographic grating polychromators are used, and some of the transfer lenses have been eliminated. The major advantages of this system are the improved stray-light rejection, wider spectral range, and decreased chromatic aberration and vignetting. As pointed out by Robinson, however, these polychromators are still quite astigmatic.[46] A nearly ideal solution to these problems would be to design two custom holographic gratings that produce the desired aberration-free spectral images. At least one manufacturer of holographic gratings has expressed confidence that such gratings could be produced, unfortunately at considerable expense.

A conventional photomultiplier detector can not be used to record the EEM because only a single signal would be registered and the spatial information present in the EEM would be lost. Rather, an imaging detector is employed. An account of the various types of imaging detectors and their characteristics is contained in Sec. B. At the time of construction of our instrument the SIT vidicon with a UV fiber-optic faceplate appeared to be the most suitable detector.

A three-dimensional display of an EEM for a mixture of perylene and anthracene obtained with the videofluorometer is shown in Figure 15. The perylene spectrum is in the lower right corner, to the red of the anthracene spectrum in the upper left: they are virtually completely separated.

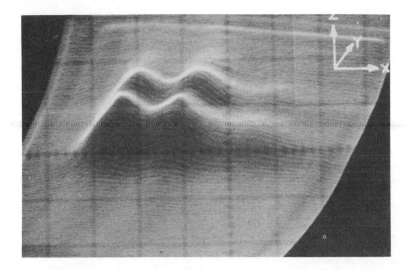

FIGURE 15. Three-dimensional display of the EEM of a mixture of perylene and anthracene obtained on the video fluorometer. Reproduced with permission from I. M. Warner, J. B. Callis, E. R. Davidson, M. Gouterman, and G. D. Christian, *Anal. Lett.* **8**, 665 (1975).

3. The Data Acquisition System

In order to perform multicomponent analyses using the EEM, as well as to provide signal averaging, difference spectra, and so forth, it is necessary to digitize and store the data. With a raster scan of 16.7 msec, and with 256 data points along each of the 256 lines, one picture element (pixel) would have to be digitized every 200 nsec. This necessitated the construction of an interface to allow efficient processing of the data without exceeding the read–write–modify cycle of the minicomputer. Details are described by Johnson *et al.*[50] Basically, the data acquisition system consists of a buffer memory and a parallel signal-processing module (pixel processor) that reduces the effective read–modify–write cycle time to 1.64 μsec. Unique features of the system are the abilities to trade off image resolution for dynamic range and to add successive images to memory for improvement of the signal-to-noise ratio. The camera and buffer memories are interfaced to a PDP 11/04 computer with 24K of memory, which is controlled with a Tektronix 4006 graphics terminal. Data may be stored on a dual floppy disk or on a videotape, or sent via phone line to a host computer (CDC 6400) for further processing (see data reduction schemes below). The analog signal may also be displayed in real time on a TV monitor.

TABLE I. SIM-I (Spectral Interpretation Model I Software)[a]

Local (PDP 11/04)		Remote (CDC 6400) data reduction
Data acquisition and reduction	Data presentation	
EEM format	Spectra	Spectral simulator
256 × 256 by 8 bit	Emission (row)	Quantitative analysis
128 × 128 by 16 bit	Excitation (column)	Least squares
64 × 64 by 32 bit	Derivative $(dI/d_{\lambda_{ex}}, dI/d_{\lambda_{em}})$	Rank annihilation
Integration	Difference $(k_1 EEM_1 - k_2 EEM_2)$	Qualitative analysis
Summation	Synchronous scan $(I_{\lambda_{ex}, \lambda_{ex}+\Delta\lambda})$	Eigenanalysis
On-target	Contour EEM	
Automatic dark-field	Three-dimensional isometric	
correction	projection	
Blank subtraction		
Difference EEM		
Chemiluminescence		
Least-squares fit		

[a] From D. W. Johnson et al.,[50] reprinted by permission.

Table I summarizes software developed for the videofluorometer, for local manipulation with the PDP 11/04 computer, as well as for multicomponent analysis using the data reduction strategies described below. Figure 16 illustrates the improvement in signal-to-noise ratio possible by utilizing the program's capabilities for summation to memory and blank subtraction. The weak perylene signal is lost in noise in the single frame, but it becomes readily apparent after summation of 512 frames and subtraction of the dark current and solvent blank. Note the target blemishes in the dark current that are subtracted out. Note also the Raman line from the solvent that is Stokes shifted from the Rayleigh-scattered light by 3000 cm^{-1}. Dark-current readings are automatically recorded after each 512 frames by use of an automatic shutter controlled by the computer.

G. DATA REDUCTION SCHEMES FOR THE EEM

While presentation of data in an EEM format provides significantly improved resolution over more conventional systems, perhaps its most valuable feature is that the matrix format allows use of various sophisticated algorithms to resolve overlapping spectra for either qualitative or quantitative analysis.

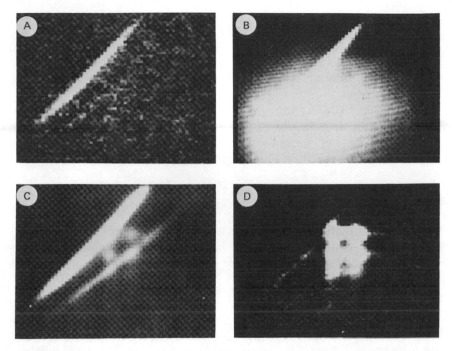

FIGURE 16. Single frame of 1×10^{-9} M perylene using 64×64 by 32 bit format. (B) Same sample after summation of 512 frames. (C) As in (B), but with subtraction of 512 dark current frames. (D) As in (C), but with blank subtraction. Reproduced with permission from D. W. Johnson, J. A. Gladden, J. B. Callis, and G. D. Christian, *Rev. Sci. Instrum.* **40**, 118 (1979).

1. Quantitative Analysis Using Simultaneous Equations and Least Squares

If one knows the qualitative composition of a solution and the spectra of the pure components in solutions of standard concentration, then it is relatively easy to determine the quantitative composition of the solution from its spectrum. Early attempts to carry out this determination were only partially successful. In these first attempts, the absorbance $A(\lambda)$ of a solution containing N components was equated at N different wavelengths to the sum $\Sigma c_i A_i(\lambda)$ of standard absorbances of each component multiplied by its concentration. This method failed because N different wavelengths could not be found such that the matrix $A_i(\lambda_j)$ labeled by i and j had a stable inverse (i.e., the simultaneous equations could not be solved meaningfully using experimental data). Sharpe[98a] attempted to overcome this difficulty by a weighted fitting of $A(\lambda)$ to $\Sigma c_i A_i(\lambda)$ at many wavelengths.

Unfortunately, Sharpe's method gave incorrect results because he was unaware of one compound present in his solution that contributed to the experimental $A(\lambda)$. Sternberg et al.[27] carried out at least a least-squares fit of $A(\lambda)$ to $c_i A_i(\lambda)$ using many λ values to solve for five components. Sustek[99] has more recently discussed the optimum choice of λ in the Sternberg scheme.

Use of the EEM data can greatly increase the stability of this approach. If one assumes that the absorbance is low so that the EEM intensity is proportional to the concentration, the intensity $I(\lambda, \mu)$ at excitation wavelength λ and emission wavelength μ should be given by

$$I(\lambda, \mu) = \sum_{i=1}^{N} c_i I_i(\lambda, \mu) \tag{5}$$

where $I_i(\lambda, \mu)$ is the corresponding intensity of standard i with concentration c_i. The matrix $I(\lambda, \mu)$ is the EEM previously described. If this is digitized at discrete wavelengths, one has a set of simultaneous equations

$$I(\lambda_j, \mu_k) = \sum_{i=1}^{N} c_i I_i(\lambda_j, \mu_k) \qquad \text{for all } j, k \tag{6}$$

As in the early attempts of Sternberg, one could try to locate a set of N different λ_j, μ_k for which the equations have a stable solution. Giering[94] has taken this approach in a recent analysis of a three-component mixture. Alternatively, one can generalize the least-squares procedure of Sternberg [see Warner et al.[100]] and determine the set of concentrations c_i that minimize

$$\sum_{jk} \left[I(\lambda_j, \mu_k) - \sum c_i I_i(\lambda_j, \mu_k) \right]^2 \tag{7}$$

Warner et al.[100] and Ho et al.[101,102] have demonstrated that this procedure can give accurate concentrations for solutions with two to six components. The procedure produces nonsensical concentrations, however, if the solution contains an important contribution to I at any of the wavelengths used in the sum that has not been included in the list of standards I_i. Methods suggested to circumvent this difficulty will be presented in Sec. G.4.

2. Component Analysis

Weber[17] noted an important fact concerning EEM matrices that is the key to more sophisticated analysis schemes. Experimentally, it has been verified that the shape of the fluorescence spectrum of a compound is independent of the exciting wavelength. The absolute intensity at low concentrations is, of course, proportional to the number of excited molecules, which, in turn, is proportional to the concentration of analyte molecules and to their absorbance at the existing wavelength. In other words,

$$I_i(\lambda_j, \mu_k) = c_i x_i(\lambda_j) y_i(\mu_k) \tag{8}$$

where $x_i(\lambda_j)$ is the standard absorption spectrum of standard i and $y_i(\mu_k)$ is its standard fluorescence spectrum.

A matrix of this form is said to have rank one. The rank of a rectangular matrix μ is defined as the dimension of the largest square submatrix with nonzero determinant. Weber noted that for a solution of several components with EEM given by $\Sigma c_i I_i$ the rank should give the number of components. Williams et al.[103] used this fact to demonstrate that the chlorella pyrenoidosa spectrum was due to at least two independent species, as the 3×3 EEM they sampled were all of rank 2.

Other types of data also yield spectral matrices whose ranks equal the number of components. For example, Wallace and Katz[104] measured a matrix $I(\lambda_j, pH_k)$ by varying the pH in a multicomponent solution. Experimental data are never perfect, so the rank of any experimental matrix is always equal to its smaller dimension. Nevertheless, with an 8×8 matrix, Wallace and Katz decided the effective rank was 3 or possibly 4.

Weber's original scheme for determining the rank based directly on its definition is quite awkward computationally. Katakis[105] suggested a better scheme based on the fact that the rank of a matrix M is the number of nonzero eigenvalues of MM^T and of M^TM. Hugus and El-Awady[106] applied this theorem to the data of Wallace and Katz and decided that the rank was 3. Bannister and Bannister[107] repeated the calculation with a different criterion for what constitutes a "zero" eigenvalue and decided that the rank was 4.

3. Eigenanalysis

Warner et al.[100] have addressed the problem of extracting qualitative information from the EEM matrix. It is highly desirable to be able to factor the observed EEM into its component parts to obtain the absorption and fluorescence spectra of each type of molecule in the solution. Unfortunately, this is, in general, impossible. Warner et al. do show, however, that the general condition for this factorization is that each molecule have a "window" in which it alone is excited and emitting. If this is the case, then it is possible to extract the complete spectrum of that molecule from the EEM. Warner et al. have discussed one algorithm for doing this for a two-component system and have examined the influence of noise on the results.

4. Partial Quantitative Analysis by Rank Annihilation

As mentioned previously, least-squares methods fail if some important constituents are present but not included in the analysis. Leggett[108] suggested that since the residual should be positive if a component is omitted from the analysis, the concentrations should be determined by a non-negative-residual least-squares approach. Warner *et al.*[100] suggested a similar scheme in which the sum of the residuals rather than their squares was minimized, subject to the condition that all residuals be positive. Both of these schemes are unsatisfactory in the sense that (a) they are extremely sensitive to the weak part of the spectrum and (b) there is no particular reason why the residual should be small if another component is present.

Ho *et al.*[101] have made a major conceptual advance in the quantitative analysis problem by noting that if the correct amount of a standard is subtracted from I, the new matrix $I - c_k I_k$ will have its rank reduced by unity. Hence, if the spectrum of this residual is examined as a function of c_k, the value of c_k is easily determined, which annihilates one additional eigenvalue. With real data, of course, the Nth eigenvalue is never zero, but there is some narrow range of c in which the Nth and $(N + 1)$st eigenvalues are of the same magnitude.

This procedure can be extended to a multicomponent analysis in which values of c are sought such that the residual

$$M = I - \sum_{k=1}^{R} c_k I_k \tag{9}$$

has rank $N - R$. Implementation of this scheme using the sum of all eigenvalues of MM^T excluding the first $N - R$ as a test function to be minimized has been implemented[102] and shown to produce reasonable results for solutions of six polynuclear aromatic hydrocarbons. For the special case $N = R$, this condition reduces to the least-squares criterion (although the calculation of optimal c_k is more complex than least squares for $N \neq R$).

H. USE OF THE EEM IN MULTICOMPONENT ANALYSIS

One potential use of the EEM is in "fingerprinting" applications, for example, in oil spill identification.* Fluorescence spectroscopy is extremely sensitive to the naturally fluorescing components of oil and to additives such

---

* The use of luminescence techniques for the characterization of oil samples and the study of weathering in oil spills is discussed in detail by Eastwood in Chapter 7 of the present volume.

as tank cleaners. Since many of the naturally fluorescent materials present, such as the polynuclear aromatic hydrocarbons (PAHs), are not particularly volatile nor soluble, fluorescence spectra are relatively stable to weathering.[109]

In 1971 Freedarde and co-workers proposed representing the EEM as contour plots, in which contours of equal fluorescence intensity are plotted against excitation and emission wavelengths to obtain characteristic finger-prints of oils (see the review by Adlard[110]). Hornig and co-workers[111,112] have pursued this approach, using their repetitively scanning spectro-fluorometer to generate the luminescence data that are taken directly into a minicomputer for correction and reduction to contour graphs.

In contrast to these applications devoted to fingerprinting, we have been interested in the use of the EEM as a quantitative tool for studies of complex mixtures. Accordingly, we have studied a number of mixtures of polynuclear aromatic hydrocarbons to test the quantitative capabilities of the videofluorometer and data reduction strategies for the EEM.

Figure 17 shows the analytical curve for 9,10-dimethylanthracene in the presence of 10^{-5} M anthracene. These compounds show highly over-lapping emission and excitation spectra. Thus, it was gratifying to find that

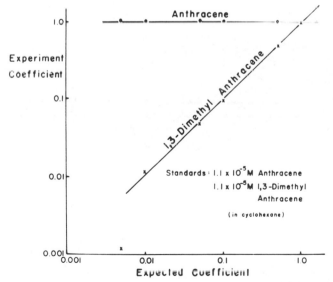

FIGURE 17. Analytical curves for the determination of 1,3-dimethylanthracene in the presence of anthracene. Reprinted with permission from D. W. Johnson, J. B. Callis, and G. D. Christian, in *Multichannel Image Detectors*, ACS Symposium Series, Vol. 102, Y. Talmi, ed. (American Chemical Society, Washington, D.C., 1979). Copyright by the American Chemical Society.

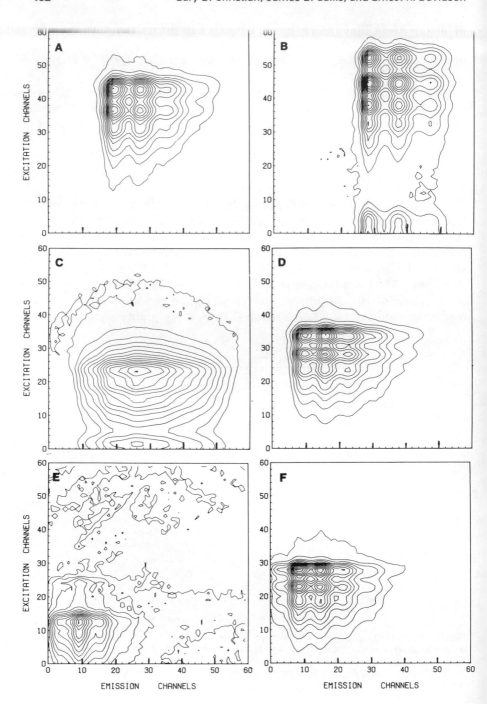

one could easily quantitate dimethylanthracene in a 100-fold excess of anthracene and that the analytical curve was quite linear. Moreover, at the spatial resolution employed in these studies there are no limitations due to pixel "cross talk," and apparently stray light in the optical system and/or "veiling glare" due to light leakage in the fiber-optic faceplate of the vidicon are not appreciable. Thus the reservations expressed by Pardue[57] and Talmi[54] do not appear to apply in terms of significantly limiting performance.

These studies and others of six-component PAH mixtures and warfarins by Johnson *et al.*[113] serve to illustrate the efficacy of simple least-squares strategies for multicomponent fluorescence analysis, where all the components are known and for which standard spectra are available. All too often, however, one is confronted with a real-world sample containing known components to be analyzed in the presence of variable background interfering unknowns. The method of rank annihilation,[101] described in Sec. G.4, is ideally suited to this problem, and Ho *et al.*[102] have extensively tested this algorithm using ten mixtures of six polynuclear aromatic hydrocarbons, each varying over wide ranges of concentration. Figure 18 shows EEMs of the six polynuclear aromatic hydrocarbon standards used as test compounds. For clarity and to show up finer details the contour levels are unevenly spaced; the nth level is $\frac{1}{2}n(n + 1)a_0$, where a_0 is the lowest level. Figure 19 is the contour plot of one of the mixture EEMs. For comparison, the conventional one-dimensional excitation and emission spectra are illustrated in Figure 20. Obviously, there is substantial spectral overlap between these components, which would make analysis by conventional techniques a difficult proposition. In spite of this overlap, Figure 18 shows that, qualitatively, the videofluorometer achieves an improved degree of separation; i.e., the components occupy distinct regions of the EEM, which is of considerable assistance for quantitative analysis. When mixtures of the six components were quantitated by the method of least squares, satisfactory answers were obtained when all six components over a wide range of concentrations were treated as knowns.[102] One component of the six was also analyzed over a range of concentrations using the concept of rank annihilation. For cases where the "null overlap" criterion of ambiguity was met[102] the average relative concentration, c(rank annihilation)$/c$(volumetric), was 1.07, with a standard deviation of 0.12. These results are truly

---

FIGURE 18. Contour maps of the spectral distributions of the components of the six-component PAH mixture (in cyclohexane). (A) 2.0×10^{-7} M perylene; (B) 2.5×10^{-6} M tetracene; (C) 9.6×10^{-6} M fluoranthene; (D) 2.8×10^{-7} dimethylanthracene; (E) 1.9×10^{-5} M chrysene; (F) 9.8×10^{-6} M anthracene. Adapted with permission from C.-N. Ho, G. D. Christian, and E. R. Davidson, *Anal. Chem.* **52**, 1071 (1980). Copyright by the American Chemical Society.

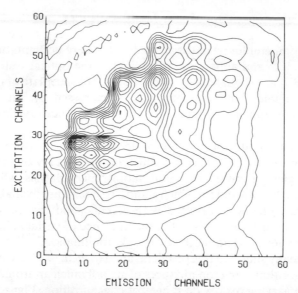

FIGURE 19. Contour maps of the six-component PAH mixture EEM. Reprinted with permission from C.-N. Ho, G. D. Christian, and E. R. Davidson, *Anal. Chem.* **52**, 1071 (1980). Copyright by the American Chemical Society.

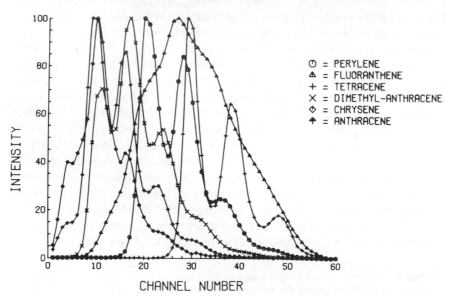

FIGURE 20. Conventional one-dimensional plot of the emission spectra of the six standards used. Each standard is normalized and taken from the most intense column of the standard EEM. Reprinted with permission from C.-N. Ho, G. D. Christian, and E. R. Davidson, *Anal. Chem.* **52**, 1071 (1980).

impressive when one considers that a single component has been satis-factorily quantitated in the presence of five other components, some of which are ten times more fluorescent! Further work has extended the concept of rank annihilation to allow simultaneous quantitation of two or more species in the presence of a large number of unknowns. Mixtures of up to 12 species have been quantitated in this fashion.[114]

As impressive as these results are, it would still be impossible to analyze quantitatively a mixture as complex as an oil sample or hydrocarbon particulate matter using the room-temperature EEM. Work of this sort will require additional resolving power. One means for accomplishing this is to improve the spectroscopic resolution by use of low-temperature sampling methods. The feasibility of this approach has been demonstrated by the excellent work of Wehry and Mamantov which is reviewed in another chapter of this monograph. The approach we have adopted is to add a third dimension to the analyses,[115] namely, temporal separation of the components via liquid chromatography (HPLC).* Hirschfeld[116] has critically appraised the multiple advantages of this approach, and we only need remind the reader of the most successful tool of this type (or of any type), the GC/MS.

A crucial element in the interface between the videofluorometer and the HPLC is the optical flow cell. Because of the sensitivity of the fluorometer to scattered light, careful attention must be paid to this inter-ference, especially if low nanogram concentrations are to be detected. Our solution to the problem is to employ a laminar-flow stream in which the sample is ensheathed in a flowing stream of solvent. With this approach one avoids scattered light and fluorescence from the windows as well as degradation of the windows from insoluble materials in the sample stream. A diagram of the laminar-flow cuvette used in these studies is presented in Figure 21. Using the videofluorometer data acquisition, one complete EEM was obtained and stored on disk every 9.4 sec. For real-time evaluation a total fluorescence chromatogram as well as two selected excitation–emission wavelength chromatograms are displayed on a graphics terminal. In addi-tion, the current EEM is displayed on a television monitor.

Some preliminary results obtained with this system illustrate capabili-ties for quantitative analysis of extremely complex mixtures. In order to ascertain the sensitivity and linearity of the system we made a series of injections of perylene ranging from 1 to 100 ng, with 9-methylanthracene (69.2 ng) as the internal standard. We found that the response was linear with the amount injected, and one nanogram was easily detected. Figure 22

---

* Luminescence methods for detection in HPLC are also discussed by Froelich and Wehry in Chapter 2 of Volume 3 of this series.

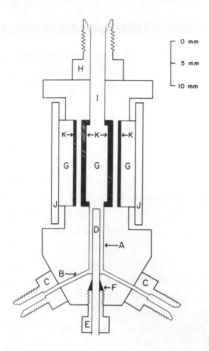

FIGURE 21. Diagram of sheath-flow cuvette. (A) Sheath alignment bore; (B) 1/32-in.-diameter sheath inlets; (C) drilled 1/16-in. Swagelok male plug; (D) 1/16-in. stainless steel tube for chromatographic effluent inlet; (E) 1/16-in. Altex male fitting; (F) 1/16-in. Teflon ferrule; (G) 1/8-in. quartz windows; (H) drilled 1/8-in. Swagelok male plug; (I) exit bore; (J) 1/16-in. stainless stell plate; and (K) 1-mm-thick Teflon gasket.

shows the total fluorescence and selected pixel chromatograms for injection of 35 ng of benzo[a]pyrene (B[a]P) and 480 ng benzo[e]pyrene (B[e]P) with 9-methylanthracene as the internal standard. Clearly, even though chromatographic resolution is not achieved, sufficient spectral resolution is obtained to provide quantitation. Indeed, 1.4 ng of B[a]P could easily be measured in the presence of 480 ng of B[e]P.

Our final fluorescence chromatogram illustrates the possibility of quantitating B[a]P in a shale oil sample. As shown in Figure 23, even with selected pixel monitoring, B[a]P is not completely separated. Even so, the EEM corresponding to the B[a]P peaks clearly shows the presence of this species. Quantitative results from this analysis are in agreement with results from a GC–MS study.

The final extension of videofluorometry to be discussed here is the analysis of fluorescent materials *in situ* on thin layer plates. Of particular importance is the use of the two-dimensional capabilities of the fluorometer to analyze a single two-dimensional chromatogram or multiple one-dimensional chromatograms simultaneously. The optical scheme for doing these measurements is given in Figure 24. Light from a xenon arc lamp is focused to infinity and passed through wavelength filters to obtain a spatially uniform beam of appropriate wavelength to excite the fluorescence of the analytes on the plate. The SIT vidicon views the plate through an appropriate wavelength filter to isolate the fluorescence of interest. In this mode, the camera serves as an electronic digital photographic plate. In order to obtain quantitative results at low concentrations we found that corrections had to be made for (a) spatial variations in excitation intensity, (b) inhomogeneities in the photo surface of the camera and (c) inhomogeneous background fluorescence on the plate. When these corrections were made, quite satisfactory quantitative results could be obtained. Figure 25 shows a fluorescent image of nine spots of tetraphenylporphine (H_2TPP) on a silica plate of various amounts ranging from 3 pg to 1.5 ng. Quantitative studies have shown that quantitation is quite feasible in the picogram region and that at least two and one half orders of magnitude range of H_2TPP can be quantitated on one plate.

Further studies have been carried out with prostaglandins derivatized by the method of Dunges.[117] Under chromatographic conditions optimized for separation of prostaglandin derivatives and their metabolites, we again

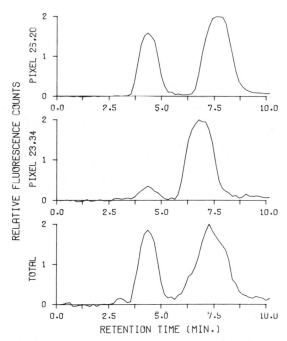

FIGURE 22. Fluorescence chromatograms of a mixture of B[a]P (35 ng) and B[e]P (480 ng) together with 9-methylanthracene (54 ng) added as an internal standard. Pixel 26,20 shows the elution of 9-methylanthracene followed by B[a]P and pixel 23,34 shows the elution of B[e]P.

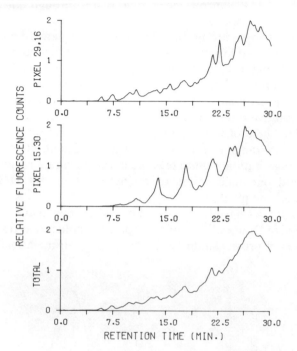

FIGURE 23. Fluorescence chromatograms for a shale oil sample. Pixel 15, 30 was chosen to give maximal response to the fluorescence of B[a]P and pixel 29, 16 was chosen in the fluorescence region for anthracenes. The peak at 22.5 min for the 29, 16 pixel corresponds to B[a]P.

obtained excellent calibration profiles extending from 100 pg to 20 ng. At present the detection limits are entirely determined by background fluorescence of the plates, which is highly nonuniform. Currently, methods are being tested for reducing the autofluorescence and subtracting it more accurately. If the background could be eliminated entirely, we calculate that subfemtogram amounts could easily be detected.

I. CONCLUSIONS

Our experience with imaging devices as fluorescence detectors leaves us with great enthusiasm for their continued use. We personally have yet to be held back by the array in any application tried, even though sensitivity, linearity, dynamic range, or channel cross talk were initially thought to be potentially severe limitations. The final obsolescence of the photomultiplier tube for specific applications must await the general availability of the large area backside-illuminated CCD array with its superb quantum efficiency and near photon-counting sensitivity. Such devices are already having a profound effect in astronomy and wherever fainter objects are being measured with ever greater precision.

Of course, the development of new strategies for data reduction must keep pace with the increasing density of data points obtained per sample. At present we have the capability to analyze two-dimensional data, as an EEM. Only recently have we begun to tackle the problem of three-dimensional data in which a series of EEMs are recorded as a function of an external variable such as chromatographic retention time, temperature, pH, etc. The efforts needed to devise strategies for analyzing these data sets will be great, but the benefits will be even greater with the quantitative analysis and analysis of ever more complex mixtures made possible.

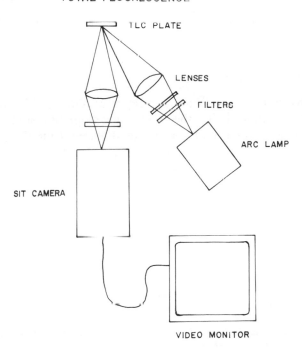

FIGURE 24. Optical system for producing a two-dimensional fluorescence image of a thin layer plate.

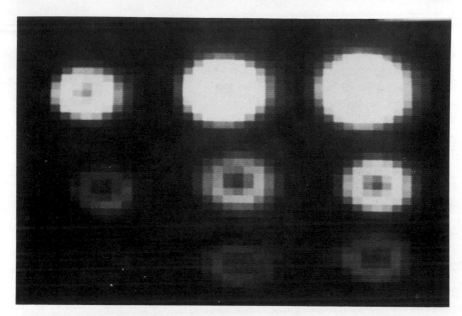

FIGURE 25. Fluorescent spots recorded with video camera for 3 pg to 1.5 ng of tetra-phenylporphine.

ACKNOWLEDGMENTS

This research was supported by NIH grants GM 22311 and GM 26935. We gratefully acknowledge the contributions of our students Isiah M. Warner, David W. Johnson, Leon W. Hershberger, Chu-Ngi Ho, and Mary Lu Gianelli.

REFERENCES

1. C. A. Parker, *Photoluminescence of Solutions with Applications to Photochemistry and Analytical Chemistry* (American Elsevier, New York, 1968).
2. J. B. Birks, *Photophysics of Aromatic Molecules* (Wiley-Interscience, London, 1970).
3. G. G. Guilbault, *Practical Fluorescence—Theory, Methods and Techniques* (Marcel Dekker, New York, 1973).
4. J. D. Winefordner, S. G. Schulman, and T. C. O'Haver, *Luminescence Spectroscopy in Analytical Chemistry* (Wiley, New York, 1972).
5. E. L. Wehry, ed., *Modern Fluorescence Spectroscopy*, Vols. 1 and 2 (Plenum Press, New York, 1976).
6. J. U. White, "Sample Optics for Increased Sensitivity in Fluorescence Spectropho-tometers," *Anal. Chem.* **48**, 2089–2092 (1976).
7. J. H. Richardson and S. M. George, "Comparison of Different Experimental Configura-tions in Pulsed Laser Molecular Fluorescence," *Anal. Chem.* **50**, 616–620 (1978).

8. G. J. Diebold and R. N. Zare, "Laser Fluorometry: Subpicogram Detection of Aflatoxins Using High Pressure Liquid Chromatography," *Science* **196**, 1439–1441 (1977).

9. T. Hirschfeld, "Optical Microscope Observation of Single Small Molecules," *Appl. Opt.* **15**, 2965–2966 (1976).

10. N. Ishibashi, T. Ogawa, T. Imasaka, and M. Kunitake, "Laser Fluorometry of Fluorescein and Riboflavin," *Anal. Chem.* **51**, 2096–2099 (1979).

11. G. L. Green and T. C. O'Haver, "Derivative Luminescence Spectrometry," *Anal. Chem.* **46**, 2191–2196 (1974).

12. T. C. O'Haver and W. M. Parks, "Selective Modulation: A New Instrumental Approach to the Fluorometric Analysis of Mixtures Without Separation," *Anal. Chem.* **46**, 1886–1894 (1974).

13. J. B. F. Lloyd, "Synchronized Excitation of Fluorescence Emission Spectra," *Nature (London) Phys. Sci.* **231**, 64–65 (1971).

14. J. B. F. Lloyd, "Partly Quenched, Synchronously Excited Fluorescence Emission Spectra in Characterization of Complex Mixtures, *Analyst* **99**, 729–738 (1974).

15. J. B. F. Lloyd and I. W. Evett, "Prediction of Peak Wavelengths and Intensities in Synchronously Excited Fluorescence Emission Spectra," *Anal. Chem.* **49**, 1710–1715 (1977).

16. P. John and I. Soutar, "Identification of Crude Oils by Synchronous Excitation Spectrofluorimetry, *Anal. Chem.* **48**, 520–524 (1976).

17. T. Vo-Dinh, "Multicomponent Analysis by Synchronous Luminescence Spectrometry," *Anal. Chem.* **50**, 396–401 (1978).

18. E. V. Shpol'skii and R. I. Personov, "Emission Spectral Analysis of Organic Compounds by Means of Spectral Lines at Low Temperatures," *Zavod. Lab.* **28**, 428–433 (1962).

19. R. Farooq and G. F. Kirkbright, "The Detection and Determination of Polynuclear Aromatic Hydrocarbons by Luminescence Spectrometry Using the Shpol'skii Effect at 77K. Part III. Luminescence Excitation Spectra," *Analyst* **101**, 566–573 (1976).

20. J. C. Brown, M. C. Edelson, and G. J. Small, "Fluorescence Line Narrowing Spectrometry in Organic Glasses Containing Parts-per-Billion Levels of Polycyclic Aromatic Hydrocarbons," *Anal. Chem.* **50**, 1394–1397 (1978).

21. R. C. Stroupe, P. Tokousbalides, R. B. Dickinson, E. L. Wehry, and G. Mamantov, "Low Temperature Fluorescence Spectrometric Determination of Polycyclic Aromatic Hydrocarbons by Matrix Isolation," *Anal. Chem.* **49**, 701–707 (1977).

22. J. L. Metzger, B. E. Smith, and B. Meyer, "Phosphorescence of Matrix-Isolated Naphthalene and Phenanthrene," *Spectrochim. Acta* **25A**, 1177–1188 (1969).

23. P. M. Roemelt, A. J. Lapen, and W. R. Seitz, "Selective Analysis of Binary Fluorophor Mixtures by Fluorescence Polarization," *Anal. Chem.* **52**, 769–771 (1980).

24. L. J. Cline Love and L. M. Upton, "Analysis of Multicomponent Fluorescent Mixtures through Temporal Resolution," *Anal. Chem.* **52**, 469–499 (1980).

25. R. B. Dickinson and E. L. Wehry, "Time-Resolved Matrix-Isolation Fluorescence Spectrometry of Mixtures of Polycyclic Aromatic Hydrocarbons," *Anal. Chem.* **51**, 778–780 (1979).

26. P. R. Bevington, *Data Reduction and Error Analysis for the Physical Sciences* (McGraw-Hill, New York, 1969).

27. J. C. Sternberg, H. S. Stillo, and R. H. Schwendeman, "Spectrophotometric Analysis of Multicomponent Systems Using the Least Squares Method in Matrix Form," *Anal. Chem.* **32**, 84–90 (1960).

28. Y. Talmi, "Applicability of TV-Type Multichannel Detectors to Spectroscopy," *Anal. Chem.* **47**, 658A–670A (1975).

29. Y. Talmi, "TV-Type Multichannel Detectors," *Anal. Chem.* **47**, 699A–709A (1979).

30. L. P. Giering and A. W. Hornig, "Total Luminescence Spectroscopy: A Powerful Technique for Mixture Analysis," *Am. Lab.* **9**, (11), 113 (1977).
31. M. Freeguard, C. G. Hatchard, and C. A. Parker, "Oil Spilt at Sea: Its Identification, Determination and Ultimate Fate," *Lab. Pract.* **20**, 35–40 (1971).
32. I. M. Warner, J. B. Callis, E. R. Davidson, and G. D. Christian, "Multicomponent Analysis in Clinical Chemistry by Use of Rapid Scanning Fluorescence Spectroscopy," *Clin. Chem.* **22**, 1483–1492 (1976).
33. I. M. Warner, J. B. Callis, E. R. Davidson, M. P. Gouterman, and G. D. Christian, "Fluorescence Analysis: A New Approach," *Anal. Lett.* **8**, 655–681 (1975).
34. Y. Talmi, ed., *Multichannel Image Detectors*, ACS Symposium Series, Vol. 102, R. F. Gould, ed. (American Chemical Society, Washington, D.C., 1979).
35. R. J. Bell, *Introductory Fourier Transform Spectroscopy* (Academic Press, New York, 1972).
36. P. R. Griffiths, *Chemical Infrared Fourier Transform Spectroscopy* (Wiley-Interscience, New York, 1975).
37. T. C. Farrar and E. D. Becker, *Pulse and Fourier Transform NMR* (Academic Press, New York, 1971).
38. P. R. Griffiths, H. J. Sloane, and R. W. Hanna, "Interferometers vs. Monochromators: Separating the Optical and Digital Advantages," *Appl. Spectrosc.* **31**, 485–495 (1977).
39. F. W. Plankey, T. H. Glen, L. P. Hart, and J. D. Wineforder, "Hadamard Spectrometer for Ultraviolet–Visible Spectrometry," *Anal. Chem.* **46**, 1000–1005 (1974).
40. H. M. Larson, R. Crosmum, and Y. Talmi, "Theoretical Comparison of Singly Multiplexed Hadamard Transform Spectrometers and Scanning Spectrometers," *Appl. Opt.* **13**, 2662–2674 (1974).
41. G. Horlick and W. K. Yuen, "Atomic Spectrochemical Measurements with a Fourier Transform Spectrometer," *Anal. Chem.* **47**, 775A–781A (1975).
42. E. M. Carlson and C. G. Enke, in *Multichannel Imaging Detectors, ACS Symposium Series*, Vol. 102, Y. Talmi, ed. (American Chemical Society, Washington, D.C., 1979), pp. 169–180.
43. R. W. Simpson and Y. Talmi, "Medium Speed Gating of ISIT Tubes," *Rev. Sci. Instrum.* **48**, 1295–1297 (1977).
44. F. Pellegrino and R. R. Alfano, in *Multichannel Imaging Detectors, ACS Symposium Series*, Vol. 102, Y. Talmi, ed. (American Chemical Society, Washington, D.C., 1979), pp. 183–198.
45. G. E. Busch and P. M. Rentzepis, "Picosecond Chemistry," *Science* **194**, 276–283 (1976).
46. G. W. Robinson, T. A. Caughey, R. A. Auerbach, and P. J. Harman, in *Multichannel Imaging Detectors, ACS Symposium Series*, Vol. 102, Y. Talmi, ed. (American Chemical Society, Washington, D.C., 1979), pp. 199–213.
47. T. M. Niemczyk, D. Ettinger, and S. G. Barnhart, "Optimization of Parameters in Photon Counting," *Anal. Chem.* **51**, 2001–2006 (1979).
48. E. J. Darland, G. E. Leroi, and C. G. Enke, "Pulse (Photon) Counting: Determination of Optimum Measurement System Parameters," *Anal. Chem.* **51**, 240–245 (1979).
49. J. A. Samson, *Techniques of Vacuum Ultraviolet Spectroscopy* Wiley, New York (1967), pp. 219–225.
50. D. W. Johnson, J. A. Gladden, J. B. Callis, and G. D. Christian, "Video Fluorometer," *Rev. Sci. Instrum.* **50**, 118–126 (1979).
51. T. Vo-Dinh, D. J. Johnson, and J. D. Winefordner, "A SIT Image Detector in Analytical Fluorescence Spectrometry," *Spectrochim. Acta* **33A**, 341–345 (1977).
52. J. A. Westphal in *Astronomical Observations with Television-Type Sensors*, J. W. Glaspey and G. A. H. Walker, eds. (UBC Press, Vancouver, 1973), pp. 127–134.

53. S. A. Colgate, E. P. Moore, and J. Colburn, "SIT Vidicon with Magnetic Intensifier for Astronomical Use," *Appl. Opt.* **14**, 1429–1436 (1975).
54. Y. Talmi, D. C. Baker, J. R. Jadamec, and W. A. Saner, "Fluorescence Spectrometry with Optoelectronic Image Detectors," *Anal. Chem.* **50**, 936A–952A (1978).
55. T. B. McCord and M. J. Frankston, "Silicon Diode Array Vidicons at the Telescope: Observational Experience," *Appl. Opt.* **14**, 1437 (1975).
56. W. B. Green, P. L. Jepsen, J. E. Kreznar, R. M. Ruiz, A. A. Schwartz, and J. B. Seidman, "Removal of Instrument Signature from Mariner 9 Television Images of Mars," *Appl. Opt.* **14**, 105–114 (1975).
57. H. L. Felkel and H. L. Pardue, "Evaluation of an Echelle Spectrometer Image Dissector System for Simultaneous Multielement Determinations by Atomic Absorption Spectroscopy," *Clin. Chem.* **24**, 602–610 (1978).
58. S. G. Vogt, R. G. Tull, and P. Kelton, "Self-Scanned Photodiode Array: High Performance Operation in High Dispersion Astronomical Spectrophotometry," *Appl. Opt.* **17**, 574–592 (1978).
59. G. Horlick, "Characteristics of Photodiode Arrays for Spectrochemical Measurements," *Appl. Spectrosc.* **30**, 113–123 (1976).
60. J. C. Geary, in *Instrumentation in Astronomy III, Proceedings of the Society of Photo-Optical Instrumentation Engineers*, Vol. 172, D. L. Crawford, ed., (SPIE Press, Bellingham, Washington, 1979), pp. 82–85.
61. W. C. Livingston, J. Harvey, C. Slaughter, and D. Trumbo, "Solar Magnetograph Employing Integrated Diode Arrays," *Appl. Opt.* **15**, 40–52 (1976).
62. R. G. Tull, J. P. Choisser, and E. H. Snow, "Self-Scanned Digicon: A Digital Image Tube for Astronomical Spectroscopy," *Appl. Opt.* **14**, 1182–1189 (1975).
63. R. G. Tull, S. S. Vogt, and P. W. Kelton, in *Instrumentation in Astronomy III, Proceedings of the Society of Photo-Optical Instrumentation Engineers*, Vol. 172, D. L. Crawford, ed. (SPIE Press, Bellingham, Washington, 1979), pp. 90–97.
64. S. B. Mende and F. H. Chaffee, "Single Electron Counting by Self-Scanning Diode Array in a Kron-Camera," *Appl. Opt.* **16**, 2698–2702 (1977).
65. S. Shectman and W. A. Hiltner, "A Photon Counting Multichannel Spectrometer," *Publ. Astron. Soc. Pac.* **88**, 960–968 (1976).
66. J. F. McNall and K. H. Nordsiek, "An Intensified Self-Scanned Array Detector System That Is Photon Noise Limited," *IAU Colloq.* **40**, 26(1)–(26)9 (1977).
67. L. Perko, J. Haas, and D. Osten in *Proceedings of the Society of Photo-Optical Instrumentation Engineers*, Vol. 116 (SPIE Press, Bellingham, Washington, 1977), pp. 56–67.
68. B. R. Sandel and A. L. Broadfoot, "Photoelectron Counting with an Image Intensifier Tube and a Self-Scanned Photodiode Array," *Appl. Opt.* **15**, 3111–3115 (1976).
69. M. Lampton, "Microchannel Plates and Their Applications to Photon Counting Image Systems," *IAU Colloq.* **40**, 32(1)–32(2) (1977).
70. B. Leskovar, "Microchannel Plates," *Phys. Today* **30** (11), 42–49 (1977).
71. J. L. Wiza, "Microchannel Plate Detectors," *Nucl. Instrum. Methods* **162**, 587–601 (1979).
72. A. J. Lieber, "Nanosecond Gating of Proximity Focused Channel Plate Intensifiers," *Rev. Sci. Instrum.* **43**, 104–108 (1972).
73. P. Schagen, in *Advances in Image Pickup and Display*, Vol. 1, B. Kazen, ed. (Academic Press, New York, 1974), pp. 1–69.
74. S. Marcus, R. Nelson, and R. Lynds, in *Instrumentation in Astronomy III, Proceedings of the Society of Photo-Optical Instrumentation Engineers*, Vol. 172, D. L. Crawford, ed. (SPIE Press, Bellingham, Washington, 1979), pp. 207–231.
75. K. L. Ratzlaff and S. L. Paul, "Characterization of a Charge Coupled Device Photoarray

as a Molecular Absorption Spectrophotometric Detector," *Appl. Spectrosc.* **33**, 240–245 (1979).

76. K. A. Hoagland, in *Low Light Devices for Science and Technology, Proccedings of the Society of Photo-Optical Instrumentation Engineers*, Vol. 78, C. Freeman, ed. (SPIE Press, Bellingham, Washington, 1976), pp. 2–9.

77. J. A. Westphal, Department of Planetary Sciences, California Institute of Technology, private communication.

78. K. W. Busch and B. Malloy, in *Multichannel Imaging Detectors, ACS Symposium Series*, Vol. 102, Y. Talmi, ed. (American Chemical Society, Washington, D.C., 1979), pp. 27–58.

79. R. P. Cooney, T. Vo-Dinh, G. Walden, and J. D. Winefordner, "Comparison of Multichannel SIT Image Vidicon and Photomultiplier Sequential Linear Scanning Systems for the Measurement of Steady-State and Transient Fluorescence of Molecules in Solution," *Anal. Chem.* **49**, 939–943 (1977).

80. R. Mathies and N-T. Yu, "Raman Spectroscopy with Intensified Vidicon Detectors: A Study of Intact Bovine Lens Proteins," *J. Raman. Spectrosc.* **7**, 349–352 (1978).

81. J. R. Jadamec, W. A. Saner, and Y. Talmi, "Optical Multichannel Analyzer for Characterization of Fluorescent Liquid Chromatographic Petroleum Fractions," *Anal. Chem.* **49**, 1316 (1977).

82. J. R. Jadamec, W. A. Saner, and R. W. Sager, in *Multichannel Imaging Detectors, ACS Symposium Series*, Vol. 102, Y. Talmi, ed. (American Chemical Society, Washington, D.C., 1979), pp. 113–133.

83. D. E. Goeringer and H. L. Pardue, "Time-Resolved Phosphorescence Spectrometry with a Silicon Intensified Target Vidicon and Regression Analysis Methods," *Anal. Chem.* **51**, 1054–1060 (1979).

84. T. Sato, Y. Suzuki, H. Kashiwagi, and M. Kakui, "Laser Radar for Remote Detection of Oil Spills," *Appl. Opt.* **17**, 3798–3803 (1978).

85. J. G. Hirschberg, A. W. Wouters, E. Kohen, C. Kohen, B. Thorell, B. Eisenberg, J. M. Salmon, and J. S. Ploem in *Multichannel Imaging Detectors, ACS Symposium Series*, Y. Talmi, ed. Vol. 102 (American Chemical Society, Washington, D.C., 1979), pp. 27–58.

86. M. M. Jotz, J. E. Gill, and D. T. Davis, "A New Multichannel Microspectrofluoro-meter," *J. Histochem. Cytochem.* **24**, 91–99 (1976).

87. C. G. Wade, R. N. Rhyne, W. H. Woodruff, D. P. Bloch, and J. C. Bartholemew, "Spectra of Cells in Flow Cytometry Using a Vidicon Detector," *J. Histochem. Cytochem.* **27**, 1049–1052 (1979).

88. M. A. Ryan, R. J. Miller, and J. D. Ingle, "Intensified Diode Array Detector for Molecular Fluorescence and Chemiluminescence," *Anal. Chem.* **50**, 1772–1777 (1978).

89. L. W. Hershberger, J. B. Callis, and G. D. Christian, unpublished observation.

90. H. W. Latz, A. H. Ullman, and J. D. Winefordner, "Limitations of Synchronous Luminescence Spectrometry in Multicomponent Analysis," *Anal. Chem.* **50**, 2148–2149 (1978).

91. G. Weber, "Enumeration of Components in Complex Systems by Fluorescence Spec-trophotometry," *Nature* **190**, 27–29 (1961).

92. I. M. Warner, G. D. Christian, E. R. Davidson, and J. B. Callis, "Analysis of Multi-component Fluorescence Data," *Anal. Chem.* **49**, 564–573 (1977).

93. L. P. Giering and A. W. Hornig, "Total Luminescence Spectroscopy. A Powerful Technique for Mixture Analysis," *Am. Lab.* **9** (11), 113–123 (1977).

94. L. P. Giering, "Multicomponent Analysis of Mixtures," *Ind. Res. Dev.* **20** (9), 134–140 (1978).

95. G. R. Haugen, B. A. Raby, and L. P. Rigdon, "On-Line Computer Control of a Luminescence Spectrometer with Off-Line Computer Data Processing and Display," *Chem. Instrum.* **6**, 205–225 (1975).

96. T. C. O'Haver, G. L. Green, and B. R. Keppler, "A Wavelength Programmed Luminescence Spectrometry," *Chem. Instrum.* **4**, 197–201 (1973).

97. J. H. Rho and J. L. Stuart, "Automated Three-Dimensional Plotter for Fluorescence Measurements," *Anal. Chem.* **50**, 620–625 (1978).

98. I. M. Warner, M. P. Fogarty, and D. C. Shelly, "Design Considerations for a Two-Dimensional Rapid Scanning Fluorimeter," *Anal. Chim. Acta* **109**, 361–372 (1979).

98a. L. H. Sharpe, Ph.D. Thesis, Michigan State University, 1957; see also Ref. 27.

99. J. Sustek, "Method for the Choice of Optimal Analytical Positions in Spectrophotometric Analysis of Multicomponent Systems," *Anal. Chem.* **46**, 1676–1678 (1974).

100. I. M. Warner, E. R. Davidson, and G. D. Christian, "Quantitative Analyses of Multicomponent Fluorescence Data by the Methods of Least Squares and Non-negative Least Sum of Errors," *Anal. Chem.* **49**, 2155–2159 (1977).

101. C. N. Ho, G. D. Christian, and E. R. Davidson, "Application of the Method of Rank Annihilation of Fluorescent Multicomponent Mixtures of Polynuclear Aromatic Hydrocarbons," *Anal. Chem.* **52**, 1071–1079 (1980).

102. C.-N. Ho, G. D. Christian, and E. R. Davidson, "Application of the Method of Rank Annihilation of Fluorescent Multicomponent Mixtures of Polynuclear Aromatic Hydrocarbons," *Anal. Chem.* **52**, 1071–1079 (1980).

103. W. P. Williams, N. R. Murtz, and E. Rabinovitch, "The Complexity of the Fluorescence of Chlorella Pyrenoidosa," *Photochem. Photobiol.* **9**, 455–469 (1969).

104. R. M. Wallace and S. M. Katz, "A Method for the Determination of Rank in the Analysis of Absorption Spectra of Multicomponent Systems," *J. Phys. Chem.* **68**, 3890–3892 (1964).

105. D. Katakis, "Matrix Rank Analysis of Spectral Data," *Anal. Chem.* **37**, 876–878 (1965).

106. F. F. Hugus and A. A. El-Awady, "The Determination of the Number of Species Present in a System: A New Matrix Rank Treatment of Spectrophotometric Data," *J. Phys. Chem.* **75**, 2954–2957 (1971).

107. W. H. Bannister and J. V. Bannister, "On Matrix Rank Analysis of Spectra of Multicomponent Mixtures," *Specialia Experientia* **30**, 972–973 (1974).

108. D. J. Leggett, "Numerical Analysis of Multicomponent Spectra," *Anal. Chem.* **45**, 276–281 (1977).

109. A. P. Bentz, "Oil Spill Identification," *Anal. Chem.* **48**, 454A–472A (1976).

110. E. R. Adlard, "Review of the Methods for the Identification of Persistent Hydrocarbon Pollutants on Seas and Beaches," *J. Inst. Petrol. London* **38**, 63–74 (1972).

111. A. W. Hornig and J. T. Brownrigg, Paper 400, 1975 Pittsburgh Conference on Analytical Chemistry and Applied Spectroscopy, Cleveland, Ohio.

112. A. W. Hornig, H. G. Eldering, and H. J. Coleman, Paper 407, 1976 Pittsburgh Conference on Analytical Chemistry and Applied Spectroscopy, Cleveland, Ohio.

113. D. W. Johnson, J. B. Callis, and G. D. Christian, in *Multichannel Image Detectors, ACS Symposium Series*, Vol. 102, Y. Talmi, ed. (American Chemical Society, Washington, D.C., 1979), pp. 97–114.

114. C.-N. Ho, Ph.D. thesis, Universtiy of Washington, 1980.

115. L. W. Hershberger, J. B. Callis, and G. D. Christian, unpublished results.

116. T. Hirshfeld, "The Hyphenated Methods," *Anal. Chem.* **52**, 297A–312A (1980).

117. W. Dunges, 4-Bromomethyl-7-Methoxycoumarin as a New Fluorescence Label for Fatty Acids," *Anal. Chem.* **49**, 442–445 (1977).

Chapter 5

Synchronous Excitation Spectroscopy

T. Vo-Dinh

A. INTRODUCTION

The increasing public awareness about the possible ecological deterioration and health hazards associated with the development of new industrial and energy processes has created more and more demanding challenges in the growing field of analytical chemistry. In the research laboratory, highly sophisticated techniques are required for providing a complete characterization of complex environmental samples. In the applied field of occupational. health protection, industrial quality control, and environmental control, simple analytical concepts and inexpensive instrumentation are also urgently needed for on-the-spot, routine, and cost-effective analyses of large numbers of samples.

Luminescence spectrometry is widely recognized as a powerful and sensitive analytical tool.[1] This technique also offers a great deal of selectivity in its ability to offer the choice of both excitation and emission wavelengths. Despite these remarkable features, the complexity of real-life samples in many practical applications is such that luminescence spectra at ambient temperatures are of little analytical value because of their relatively featureless appearance. One technique that could offer an improved selectivity for luminescence analysis, without having to sacrifice simplicity, is the synchronous excitation technique. This chapter describes the methodology

T. VO-DINH • Health and Safety Research Division, Oak Ridge National Laboratory, Oak Ridge, Tennessee 37830

and presents the proportion of this important technique; it also discusses the advantages as well as the limitations of the method and includes some representative applications.

B. METHODOLOGY AND INSTRUMENTATION

1. Basic Principles

In conventional luminescence spectroscopy two basic types of spectra are usually measured. When a sample is excited at a fixed wavelength λ_{ex}, an emission spectrum is produced by recording the emission intensity as a function of the emission wavelength λ_{em}. Conversely, an excitation spectrum may be obtained when λ_{ex} is scanned while the observation is conducted at a fixed λ_{em}. Recently, a third method has been suggested that consists of recording the luminescence signal while simultaneously (or synchronously) scanning both λ_{ex} and λ_{em}.[2,3] This technique is called "synchronous excitation luminescence spectroscopy," or more simply "synchronous luminescence" (SL). The synchronous excitation method can be applied to fluorescence (synchronous fluorometry) as well as phosphorescence (synchronous phosphorimetry) analysis. Although the general term SL is applied to most applications, all the discussions, except those in Sec. C.3, deal with synchronous fluorometry.

In the SL method a constant wavelength interval $\Delta\lambda$ is maintained between the excitation and emission wavelengths throughout the measurement. The intensity of the synchronous signal was shown to have the following form[3]:

$$I_{SL}(\lambda_{ex}, \lambda_{em}) = KcbE_X(\lambda_{ex})E_M(\lambda_{em}) \tag{1}$$

with

$$\Delta\lambda = \lambda_{em} - \lambda_{ex} = \text{const} \tag{2}$$

where c is the concentration of the analyte, b is the thickness of the sample, $E_X(\lambda_{ex})$ is the excitation spectrum, $E_M(\lambda_{em})$ is the emission spectrum, and K is a constant that includes the usual "instrumental geometry factor" and related parameters.

Using Equation (2), Equation (1) can be explicitly expressed either as a function of λ_{em} or as a function of λ_{ex}:

$$I_{SL} = KcbE_X(\lambda_{em} - \Delta\lambda)E_M(\lambda_{em}) \tag{3}$$

$$I_{SL} = KcbE_X(\lambda_{ex})E_M(\lambda_{ex} + \Delta\lambda) \tag{4}$$

From Equation (4), it can be seen that the synchronous signal usually referred to as "a synchronously excited emission signal" could also be considered as an excitation spectrum with a synchronously scanned emission wavelength.

Equation (1) shows that the synchronous spectrum depends explicitly upon both the emission and excitation spectra. The intensity profile of the synchronous spectrum contains, therefore, the combined properties of both the emission and excitation spectra. In other words, the SL concept takes advantage of the spectral characteristics of the absorption properties as well as the emission properties of a given compound. It is this apparently trivial feature that results in the improved selectivity of the SL technique for multicomponent analysis.

Figure 1 illustrates the basic effect of the SL technique in fluorescence spectroscopy. The fluorescence emission ($\lambda_{ex} = 407$ nm) and excitation

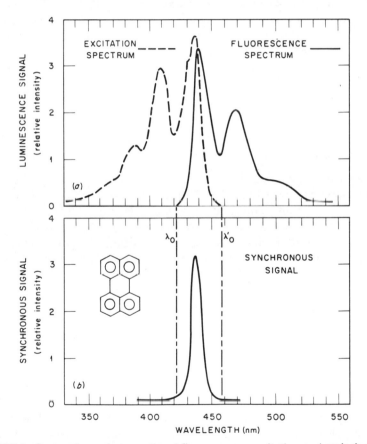

FIGURE 1. Comparison of conventional fluorescence excitation and emission spectra for perylene in ethanol (a) with synchronous fluorescence spectrum ($\Delta\lambda = 3$ nm) of perylene in ethanol (b). Reprinted with permission from Vo-Dinh *et al.*[4] Copyright by the American Chemical Society.

(λ_{em} = 470 nm) spectra of perylene in ethanol are shown in Figure 1a. The small shift of about 3 nm between the 0–0 bands in the excitation and emission spectra is often referred to as the Stokes shift $\delta\lambda_s$. The fluorescence emission spectrum exhibits three main bands at about 440, 470, and 505 nm.[4] This spectral profile remains unchanged when other excitation wavelengths are used. However, if the synchronous technique is applied, using $\Delta\lambda$ = 3 nm in order to match the Stokes shift, only one single emission band is obtained (Figure 1b).

2. Analytical Characteristics and Advantages

The major problem area in luminescence spectrometry of multi-component samples, especially at ambient temperatures, is the selectivity. The two possible causes for this limitation are the superimposition of numerous spectra from different species and/or the inherently featureless nature of the individual spectra. The synchronous excitation technique can be shown to decrease the adverse effect due to these two sources of spectral diffuseness. This technique can produce:

(a) a bandwidth-narrowing effect,
(b) a spectral profile simplification, and
(c) a reduction of spectral overlap as a result of confinement of the wavelength region occupied by each individual spectrum.

Of particular interest is the fact that the bandwidth of the synchronous signal is usually narrower than the width of the corresponding band in the conventional spectrum. This feature results essentially from the multi-plicative effect of the two functions E_M and E_X increasing and/or decreasing simultaneously while both λ_{em} and λ_{ex} are scanned. Figure 2 illustrates this bandwidth-narrowing effect with the fixed-excitation (λ_{ex} = 442 nm) and synchronous ($\Delta\lambda$ = 3 nm) fluorescence spectra of tetracene.

The synchronous technique is attractive and useful even when a compound exhibits no spectral structure in either the emission or excitation spectra. In this case, $\Delta\lambda$ can be chosen such that the product $E_M E_X$ applies to the spectral region near the sharp edges of these two functions. The resulting synchronous signal can exhibit a much narrower band. Figure 3 illustrates this effect for o-cresol. The remarkable narrowing of the bandwidth in the synchronous signal ($\Delta\lambda$ = 3 nm) underscores the analytical usefulness of the technique.

The synchronous concept also enhances the chance of observing a structured spectrum. A conventional luminescence spectrum exhibits a resolved structure only when the monitored spectrum (E_M or E_X) consists itself of some narrow-band features. But in the synchronous method a

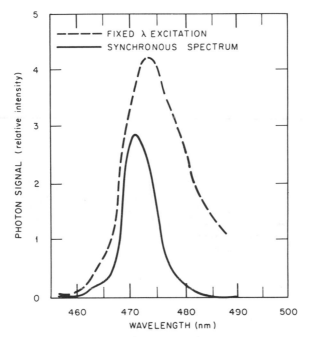

FIGURE 2. Comparison of fixed-wavelength excitation and synchronous fluorescence spectra for tetracene. Note the narrower band obtained when synchronous excitation is employed. Reprinted with permission from Vo-Dinh.[3] Copyright by the American Chemical Society.

narrow peak can be observed even when either the excitation or the emission spectrum has a resolved structure within the spectral range under study. For example, a compound that has a diffuse emission spectrum but exhibits a structured excitation spectrum can produce a structured synchronous signal profile.

Spectral simplification is another attribute of the synchronous technique. During the synchronous scanning procedure the emission of each individual compound is simplified to a single band or a few narrow peaks. Most of the information contained in the other spectral bands is lost. For the photophysical spectroscopist, this loss of information would certainly be a severe disadvantage, because all the spectral details in the entire spectrum are of utmost importance. For the practical analytical researcher the properties of the entire spectrum might not be of crucial importance, because only one or a few emission bands are generally selected for his analytical work. Not only are all the other emission bands often not considered, but their presence could interfere with the emissions from the other species. The

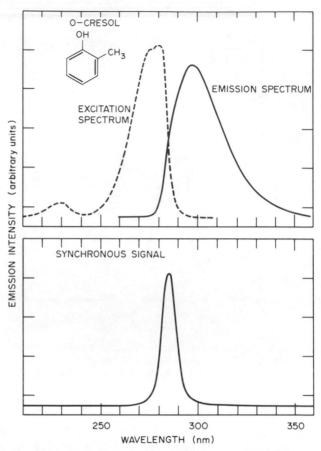

FIGURE 3. Comparison of conventional fluorescence excitation and emission spectra (top) with synchronous fluorescence spectrum (bottom) for o-cresol. Reprinted with permission from Vo-Dinh *et al.* Copyright by Ann Arbor Scientific.

synchronous excitation approach simplifies the emission of each individual component in order to reveal a more resolved structure from the composite system.

In conjunction with the spectral simplification process the ability to confine the individual emissions within specific small spectral ranges is another outstanding feature of the synchronous technique. Since the emission function E_M is limited in the short-wavelength range and the excitation function E_X is limited in the long-wavelength range, the exclusive nature of the multiplication of E_M and E_X produces a synchronous signal limited within a specific spectral interval. This spectral range is defined by the parameter $\Delta\lambda$ and the Stokes shift $\delta\lambda_s$ of the compound. It is experimentally possible to select the spectral bandwidth of the synchronous signal by using appropriate values of $\Delta\lambda$ and $\delta\lambda_s$. Since the Stokes shift is the consequence of the difference in configuration of a molecule in its ground and excited singlet states in a given medium, a variation of the solvent polarity can induce a change in $\delta\lambda_s$. This important feature has so far not been fully investigated and has yet to be experimentally exploited. The reduction of the spectral range is clearly illustrated in Figures 1 and 3. The conventional spectrum of perylene covers a large spectral range between 430 and 530 nm, whereas the synchronous signal has a half-width of only about 10 nm (Figure 1). Note also the broad bandwidth (40 nm) of the conventional fluorescence of o-cresol and the much narrower bandwidth (10 nm) of the synchronous peak (Figure 3).

The remarkable advantage of this spectral reduction effect is fully exploited only in the analysis of complex mixtures. Figure 4 shows the fluorescence spectra of a mixture of five polyaromatic hydrocarbons (PAH). This mixture consists of anthracene, naphthalene, perylene, phenanthrene, and tetracene. Although the conventional spectrum (Figure 4a) exhibits several relatively resolved bands, the severe spectral overlap renders any band assignment ambiguous. Also, the simple rule correlating the spectral position to the ring size of the PAH compound cannot be efficiently applied because of this spectral superposition. With the synchronous technique, however, this simple rule could provide the means for the determination of the presence of a given compound (or group of compounds) in a mixture by a simple "screening" type of measurement (Figure 4b). The next section will discuss this feature in more detail.

3. The Choice of $\Delta\lambda$

The synchronous technique was initially suggested and used for obtaining qualitative spectral fingerprints of forensic samples.[2] Because these real-life samples are extremely complex, their luminescence spectra are mostly diffuse and structureless. Even the use of various excitation wavelengths cannot satisfactorily resolve the diffuse structure. Their SL spectra, however, could exhibit some useful features. In these applications, the choice of $\Delta\lambda$ is usually empirical and must be determined for each application.[5-7] In general, the recommended $\Delta\lambda$ value is the one that produces the most resolved spectral structure. The $\Delta\lambda$ parameter can have values ranging from 15 to 100 nm. For example, several $\Delta\lambda$ values between 15 and 100 nm were used to characterize rubber materials in tires.[6] Other values of $\Delta\lambda$ between 20 and 40 nm were also found to be appropriate for identifying crude oils from different sources.[7]

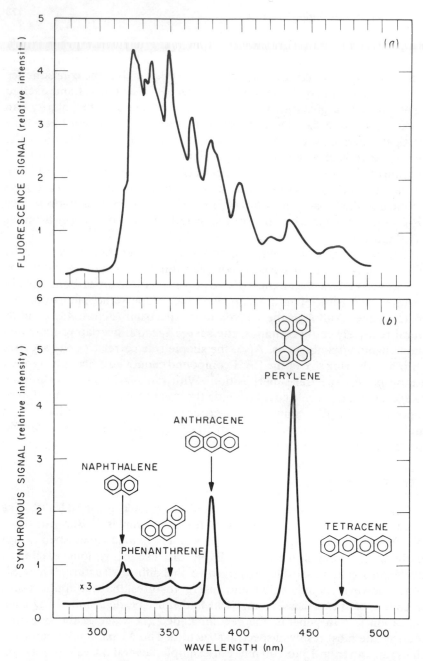

FIGURE 4. The synchronous excitation approach to multicomponent analysis. (a) Fixed-excitation fluorescence spectrum of a synthetic five-component mixture; (b) synchronous fluorescence spectrum of the same sample. Reprinted with permission from Vo-Dinh.[3] Copyright by the American Chemical Society.

Recently, the concept of synchronous excitation has been further developed and extended to multicomponent analysis.[3-5,8-10] In this approach, the goal is to develop a simple method for qualitative and quantitative trace analysis. The purpose is to investigate how the synchronous technique can be applied to obtain not only spectral signatures from complex samples but also compound-specific information of analytical interest. In multicomponent analysis, the entirely empirical approach for the choice of $\Delta\lambda$ is no longer desirable because the aim is not only to produce a few more peaks in the spectrum but also to provide a reliable means for correlating each emission band to certain known and invariant photophysical parameters of the corresponding species.

Of particular interest are the SL characteristics of the important family of polycyclic compounds. The use of $\Delta\lambda$ equal to the small Stokes shift $\delta\lambda_s$ of these compounds underscores the simple correlation between the intensity profile of the SL spectrum and the composition of the sample. The main relationship between the size of a PAH compound and its fluorescence spectrum is reflected by the dependence of its 0–0 band energy upon its number of aromatic rings. The vibronic structure of its fluorescence spectrum is much less compound specific because most of PAH spectra exhibit a principal series of vibrational bands of diminishing intensity that are evenly spaced at frequency intervals of 1400 cm^{-1} owing to the dominant C–C vibrational modes. On the other hand, the wavelength position of the 0–0 band of a high-ring-number linear cyclic compound generally occurs at a longer wavelength than that of a lower-ring-number compound. Spectroscopic data pertaining to the 0–0 band and the $\delta\lambda_s$ value of many compounds can be found in or derived from published works or a spectral handbook.[11]

Most PAH molecules having two to five rings usually exhibit a fluorescence 0–0 band between 300 and 500 nm and have $\delta\lambda_s$ values between 3 and 5 nm. Since the emission bandwidths are much broader, usually on the order of 10–15 nm, the use of one value of $\Delta\lambda (=3 \text{ nm})$ is appropriate for most of these polyaromatic compounds. It is important to note that 3 nm is only an average value. Depending on the specific compounds present in a multicomponent sample or the ability of the monochromators to suppress scattered light, slightly different values of $\Delta\lambda$ (between 2 and 5 nm) can be used. Whenever possible, it is advantageous to match $\Delta\lambda$ with $\delta\lambda_s$. This situation provides a single peak having the most intense synchronous signal and the narrowest half-width.[3] It is noteworthy to mention that the perfect matching of $\Delta\lambda$ and $\delta\lambda_s$ is not always possible for all the compounds in a mixture because the $\delta\lambda_s$ values for different compounds can differ slightly from one another. For those compounds with no perfect matching, the synchronous signal exhibits a typical double structure having two closely spaced peaks. Figure 5 shows the structure of the synchronous fluorescence signal of a

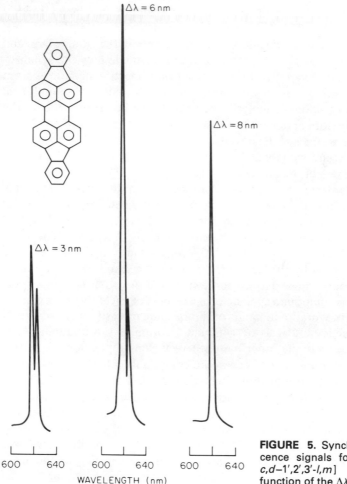

FIGURE 5. Synchronous fluorescence signals for diindeno[1,2,3-c,d–1',2',3'-l,m] perylene as a function of the $\Delta\lambda$ value.

large molecule, diindeno[1,2,3-c,d–1',2',3'-l,m]perylene, using three different $\Delta\lambda$ values. Exceptions to the general rule that requires matching between $\lambda\lambda$ and $\delta\lambda_s$ are the situations where the 0–0 bands in absorption and/or emission are too weak or where the scattered light is too intense with the use of a small $\Delta\lambda$ value.

4. Spectral Interferences

Besides the first type of limitation, e.g., featureless structure and spectral overlap, other types of phenomena such as inner- and postfilter effects, quenching, and energy-transfer processes can also limit the applicability of luminescence spectrometry under certain conditions. This section deals with the limitations caused by the second type of problem. Of special interest is the question as to how, when, and in what form those problems appear in SL spectra. A misunderstanding of these limitations may produce ambiguous results.

First, it is important to recognize that the filter effect is a serious limitation in multicomponent luminescence analysis by any method, either conventional or synchronous. The total absorbance of the sample must be less than 10^{-2} in all spectral regions. As more and more components are present, the chances of spectral overlap increase, resulting in a high total absorbance value. Nonlinearity of the analytical curves,[4] distortion of the SL profile,[7] or decrease in the intensity of a certain number of emission bands are commonly observed at high concentrations.[10,12,13] Great caution must be used such that solutions are sufficiently dilute to avoid the artifacts caused by these spectral interferences. Quantification is possible at concentrations corresponding to the linear portions of the analytical curves. Examples of analytical curves for SL signals obtained from shale oil water are shown in Figure 6.

At high concentrations, quenching and energy-transfer processes also produce spectral distortions. The origins of these phenomena can be either of a collisional or a long-range type. The collisional processes (also called

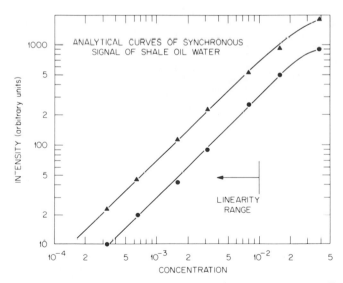

FIGURE 6. Analytical calibration curves for the two principal synchronous fluorescence peaks (▲, 326 nm; ●, 330 nm) from a shale oil water sample.

diffusion-controlled processes) can be avoided by using viscous media such as polymer matrices or low-temperature solvents.[14,15] More research and development work remains to be conducted in this area for SL applications. Energy-transfer processes generally cause the migration of excitation energies from compounds of high energies to compounds of lower energies. A shift of the spectral profile towards longer wavelengths is, therefore, the general trend observed.[7] With regard to these energy-transfer processes, the usually small $\Delta\lambda$ value between the excitation and observation wavelengths should decrease the possibility of energy transfer in SL.

In practical analyses, the inner-filter effect is the main problem to be considered because the requirement of low absorbances is usually adequate to ensure that neither energy transfer nor collisional quenching will occur.

Figure 7 illustrates the type of spectral distortions often observed in SL spectra at high concentrations.[10] Note the drastic change in intensity

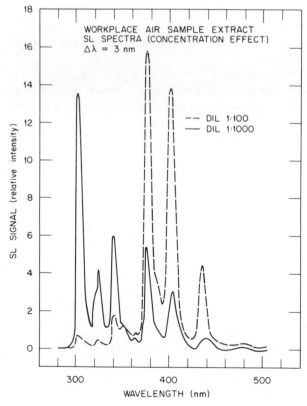

FIGURE 7. Variation with concentration of synchronous fluorescence spectrum of a complex sample.

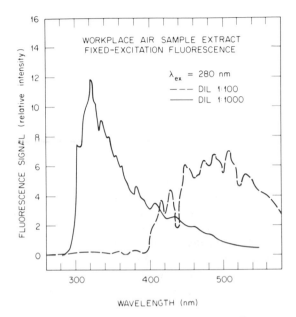

FIGURE 8. Variation with concentration of fixed-excitation fluorescence spectrum of the complex sample whose synchronous spectra are shown in Figure 7.

distribution of the SL spectra from a workplace aerosol extract that has been subjected to two different dilutions. In this example, a simple dilution of the sample can suppress these spectral variations. All the spectra obtained with the aerosol extract samples that are more than 1000-fold diluted retain the same profile. It is important to emphasize that the spectral distortions due to high concentrations also occur in conventional spectroscopy. Figure 8 shows that this type of spectral interference is not specific to the SL method but rather a common problem inherent to luminescence spectroscopy of liquid solutions. The extent to which the sample concentration affects the conventional spectra is underscored by the almost complete suppression of the emission at short wavelengths between 300 and 400 nm.

The synchronous excitation approach can improve the selectivity of luminescence analysis by decreasing the effect of the first type of problem, e.g., the spectral overlap and the broad structures, but it cannot suppress the second type of problem (quenching, inner-filter effects, etc.). This situation is to be expected because the first kind of problem is related to the nature of the data, whereas the second kind of problem is related to the actual photophysical processes occurring within the sample. These problems can only be modified by selecting suitable working conditions and should not

depend upon the method of recording the data (conventional or synchronous spectrometry).

Another type of problem that is often underestimated is scattered light.
There are three types of scattering phenomena that can interfere with a
luminescence measurement. Two types of scatter peaks, Raleigh and
Tyndall scatter, have the same wavelength as the excitation light. Tyndall
scatter arises from reflection from particulate matter present in a sample,
whereas Raleigh scatter originates from an interaction between the excitation light and the solvent molecule. The third type, Raman scatter, also
involves an interaction between the excitation light and the solvent but
induces a red shift. Tyndall and Raleigh scattering may interfere in SL
measurements because the small interval between λ_{ex} and λ_{em} can produce
significant stray light. To eliminate this effect, the slit width must be no
greater than half of the selected $\Delta\lambda$ value. For example, if $\Delta\lambda$ is chosen to be
equal to 3 nm, the slit widths must be equal to or less than 1.5 nm. One must
recognize that the use of such a small slit width would decrease the light
throughout and, consequently, the sensitivity. Nevertheless, it is also
important to consider the trade-off between spectral resolution and sensitivity. Since most of the SL peaks have bandwidths on the order of 4 nm, the
use of broad slit widths (>2 nm) would cause the resolution to deteriorate.

Raman scatter is usually not a major problem in SL measurements.
Raman peaks can be predicted to appear at specific wavelengths and can be
used as references for the purity of the solvents.[16]

5. Two-Dimensional Representation of the Synchronous Excitation Concept

The SL technique can also be schematically visualized in a two-
dimensional representation. This representation is described in detail in
Chapter 4 of this book, which deals with total luminescence (TL) spectroscopy. In the TL representation the three-dimensional hypersurface that
corresponds to the combined excitation and emission functions is projected
onto the two-dimensional plane defined by the coordinates λ_{em} and λ_{ex}. This
approach produces an isocontour spectrogram in the same manner as in
topographic maps.

Figure 9 illustrates the concept of synchronous scanning in the two-
dimensional TL representation. An SL spectrum corresponds to the
luminescence intensity measured along the line defined by the equation
$\lambda_{em} = \lambda_{ex} + \Delta\lambda$. Parallel to this scanning line is the diagonal which, by
definition ($\lambda_{em} = \lambda_{ex}$), corresponds to Rayleigh scattering. Note the value of
the Stokes shift graphically defined in Figure 9. The total luminescence
spectrum of the hypothetical compound shown in Figure 9 consists of a

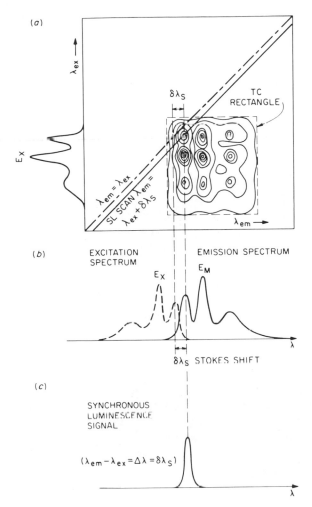

FIGURE 9. Definition of the relationships between the total luminescence or two-dimensional contour plot and synchronous fluorescence techniques. The fluorescence and emission spectra of the particular compound in question are shown. The Stokes shift is represented by $\delta\lambda_s$. (a) Total luminescence spectrogram, (b) conventional excitation and emission spectra, (c) synchronous luminescence spectrum.

multicontour spectrogram enclosed within the boundaries of a rectangle referred to as the total contour (TC) rectangle. A mixture of several PAH compounds having a given $\delta\lambda_s$ value therefore exhibits TL spectrograms aligned along the diagonal scatter line. Figure 10 shows the TL spectrograms for four compounds labeled A, B, C, and D schematically represented by their total contour rectangles. Figure 10 shows that one single synchronous

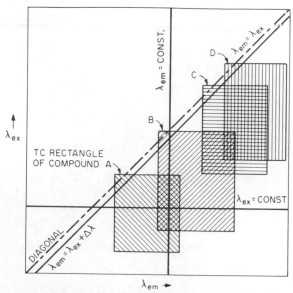

FIGURE 10. Schematic representation of the location of the rectangular contours in the total luminescence spectra of four hypothetical compounds A, B, C, and D. The diagonal defines the Rayleigh scatter line ($\lambda_{em} = \lambda_{ex}$), as in Figure 9. Note that the choice of the proper value for $\Delta\lambda$ allows one synchronous scan to pass through the rectangular contour region for each compound.

scan (with $\Delta\lambda = \delta\lambda_s$) can consecutively "slice through" all the TL spectra. Although the SL scanning method discards the TL information outside the scanning line, it is a very simple and effective means for obtaining data for several compounds in one measurement.[17]

6. Instrumentation

One of the most attractive features of the synchronous excitation technique is the simplicity of instrumentation. Any commercial or laboratory-constructed luminescence spectrometer that has an excitation and an emission monochromator can be employed for SL measurements. Several instruments with the interlocking capability of the two monochromators as a means of conducting absorption measurements are commercially available. Even with a spectrometer without the interlocking feature, a simple and inexpensive switch can be added for synchronous scanning.

In recent published works, several devices have been suggested as other means for obtaining SL spectra. These suggested devices include a contour plotter[18,19] and a videospectrometer with a diagonal scanning capability across the spectral matrix.[20] Whereas these instruments could be useful for certain research and development work by providing different tools to evaluate the usefulness of the SL technique, the basic luminescence spectrometer is the simplest and most suitable for SL analyses, especially for routine applications.

C. ANALYTICAL APPLICATIONS

1. A Fingerprinting Technique

Lloyd has reported an extensive number of studies in which the SL method was used to obtain spectral fingerprints of complex forensic samples.[21] For these samples, the conventional fluorometric method yields little information because the fixed-excitation spectra usually show diffuse and featureless structures. On the other hand, the SL technique using various $\Delta\lambda$ values between 15 and 100 nm produces spectra that contain a few peaks which provide unique fingerprints for each sample. In Lloyd's studies tires of various origins were characterized by the spectral features of their SL profiles. The variations in the spectra were used to reveal aging effects on the tire treads. Using selected $\Delta\lambda$ values, the SL measurements allowed a clear distinction between rubbers of the tire sidewalls and the tread. Following a chemical extraction, the fluorescent contact traces transferred from various rubber articles exhibited typical SL spectra indicating the presence of extender and process oils.[21] John and Soutar also used the SL technique to identify various crude oils.[7] Various studies demonstrate that the SL technique shows great promise as a diagnostic tool for the identification of oil spillages in the marine environment.[7,22] Two examples of SL spectra (a tire print extract and Ecofisk oil) are shown in Figure 11.

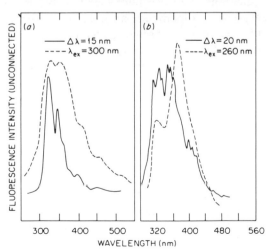

FIGURE 11. The synchronous luminescence technique used as a fingerprinting tool. (a) Synchronous (——) and fixed-excitation (– – –) spectra of an extract from a tire print. From Lloyd[6]; reproduced by permission of The Chemical Society (London) (b) Synchronous (——) and fixed-excitation (– – –) spectra of Ecofisk crude oil. From John and Soutar[7]; reproduced by permission of the American Chemical Society.

✳ 2. Quantitative Multicomponent Analysis

Recently, the synchronous excitation concept has been further developed to provide not only spectral signatures of complex samples but also qualitative and quantitative data for multicomponent analysis.[3,4,8–10,23]

An example of a practical application of the SL method in pollution analysis is the identification of the naphthalene analogs in a coal gasifier byproduct water by synchronous fluorometry.[4] The SL technique has been shown to be useful for the determination of cresol isomers in this wastewater.[23] The content of phenol and o-, m- and p-cresol in the wastewater cannot be determined by the fixed-excitation method because the conventional fluorescence spectra of these compounds are all quite featureless and similar. As previously shown in Figure 3 for o-cresol, the remarkable narrowing of the emission band in the SL spectrum greatly improves the specificity of the fluorescence assay. In the analysis of this wastewater, this technique has been used in conjunction with gas chromatography to yield a complete quantitative analysis of all three cresol isomers.

Recently, the SL technique has been used to identify and quantify polynuclear aromatic (PNA) compounds in an atmospheric aerosol extract from the Source Assessment Sampling System (SASS). The SASS scheme with its associated analytical procedures is a sampling and analytical approach developed by the U.S. Environmental Protection Agency (EPA) for conducting environmental source assessments of the feed, product, and waste streams associated with industrial and energy-related processes.[24] In the analysis of this aerosol extract, the SL method ($\Delta\lambda = 3$ nm) can identify and quantify up to 13 major and trace PNA compounds, including anthracene, benzo[b]fluorene, benzo[a]pyrene (B[a]P), benzo[e]pyrene, chrysene, dibenz[a,h]anthracene, dibenzothiophene, fluoranthene, fluorene, phenanthrene, perylene, pyrene, and tetracene (Table I).[25] Figure 12 shows the conventional and synchronous fluorescence spectra of this sample. The presence of eight major PNA components is clearly revealed in one single SL measurement (Figure 12b).

Figure 13 illustrates another example of practical application of SL. This figure shows the effectiveness of the SL method in purity control procedures for various phenanthrene samples. All the SL measurements were conducted with $\Delta\lambda = 3$ nm. The peaks labeled P are from pure phenanthrene, whereas the other peaks indicate the presence of phenanthrene photoproducts or unidentified contaminants. The SL spectrum of freshly prepared phenanthrene (commercial grade) exhibits a phenanthrene to impurity ratio higher than that of an old sample, but still contains an appreciable amount of impurity (Figures 13a and 13b). Figure 13c indicates the effective removal of most of the contaminants by a quick thin-layer

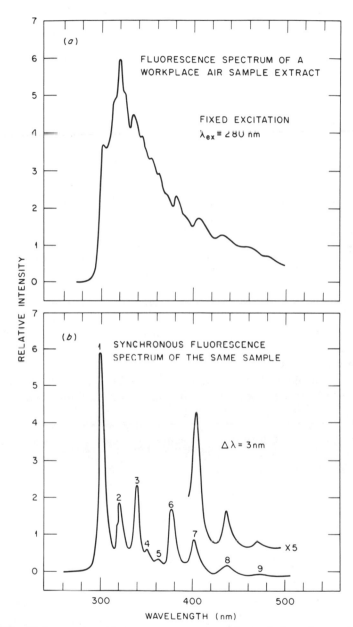

FIGURE 12. (A) Conventional fluorescence spectrum of a workplace air sample extract. (b) Synchronous spectrum of the same sample. Peak assignments: 1, fluorene; 2, naphthalene + acenaphthene; 3, benzo[b]fluorene; 4 and 5, chrysene; 6, anthracene; 7, benzo[a]pyrene; 8, perylene.

TABLE I. Quantitative Analysis of a Workplace Aerosol
Extract by Synchronous Fluorescence Spectroscopy

Compound	Concentration[a] (M)
Anthracene	7.7×10^{-7}
Benzo[a]pyrene	6.0×10^{-7}
Benzo[e]pyrene	$<6.0 \times 10^{-6}$
Benzo[b]fluorene	3.3×10^{-7}
Chrysene	3.2×10^{-6}
Dibenz[a,h]anthracene	$<5.0 \times 10^{-7}$
Dibenzothiophene	$<4.0 \times 10^{-7}$
Fluoranthene	1.9×10^{-6}
Fluorene	1.2×10^{-7}
Phenanthrene	$<2.0 \times 10^{-5}$
Perylene	1.0×10^{-8}
Pyrene	3.0×10^{-6}
Tetracene	1.9×10^{-8}

[a] Data obtained with a 1000-fold diluted sample.

chromatographic separation. This example suggests that the SL technique
can offer a rapid and simple tool that is much needed in the areas of quality
control and industrial process control, where simple and quick measurements must be routinely conducted.

Another practical application of SL is the fluorescence analysis of
tracers in hydrology. Many compounds such as fluorescein, eosin, sulfor-
hodamine G, rhodamine B, and WT are used as fluorescent tracers in

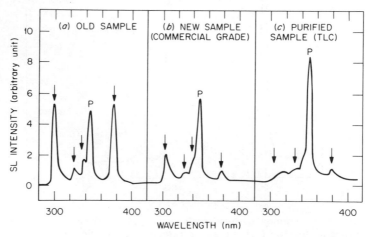

FIGURE 13. Comparison of synchronous fluorescence spectra of various impure
phenanthrene samples. Arrows denote impurities.

monitoring the course of underground waters. André and his co-workers have demonstrated that the synchronous technique greatly improves the determination of these tracers in aqueous samples.[26] First, the SL method enhances the sensitivity and reduces experimental errors. Secondly, spectral interference from the scattered light at the excitation wavelength is also decreased. Finally, whereas the measurement of a blank is necessary with the conventional fixed-excitation technique, it is not needed with the SL method, even in quantitative analyses, since the blank value may be obtained by direct extrapolation of the data.[26] In this context, the synchronous excitation method is particularly suitable for monitoring the concentration of fluorescent tracers in hydrological studies because it is often impossible to conduct blank measurements in the field.

Recently, SL has been used in several other applications of biological and environmental interest. These applications include the identification and quantification of the carcinogenic polycyclic aromatic hydrocarbon dibenz[a,h]anthracene in extracts of cigarette smoke[27] and the detection of the binding of B[a]P to epidermal DNA and RNA.[28] Binding of carcinogenic B[a]P to epidermal DNA occurs following the *in vivo* application of crude oils to animal skin. In conjunction with the photon-counting technique, the SL method was used to enhance the sensitivity of the fluorescence assay to the point at which levels of B[a]P binding can be detected in the DNA isolated from the epidermia of a single mouse. Rahn and co-workers suggested that this nonradiative assay be used as a short-term biophysical assay for the determination of the biological hazards associated with synthetic fuels.[28]

In some cases, enhancement in selectivity is achieved by taking derivatives of the luminescence signal output. This method, referred to as derivative luminescence spectroscopy, has been described by O'Haver.[29] It consists of measuring the first, second, or higher derivative of a spectrum with respect to wavelength and recording this derivative rather than the spectrum itself. This technique is used to enhance minor spectral features superimposed on a broad and intense background. In the second-derivative (d^2) mode, the curvature of a peak is measured rather than the intensity of the peak itself. Broad bands tend to be eliminated in the d^2 recording mode while sharp spectral features are intensified. Figure 14 illustrates the application of the d^2 method to SL spectroscopy. In Figure 14a, the sample is a synthetic reference mixture analyzed by several laboratories in a round robin in order to verify the use of the Level 1 scheme, the first step of a phased approach developed by the EPA for the sampling and chemical analysis of organic constituents in industrial effluents. Table II lists the measured quantities and the actual amounts of the individual compounds present in the sample. The effectiveness of the d^2 SL analysis is

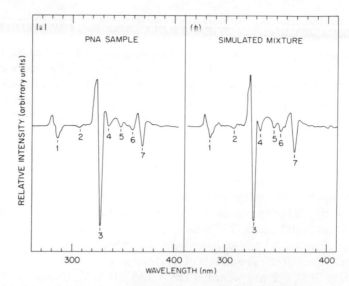

FIGURE 14. Second derivative synchronous fluorescence spectra ($\Delta\lambda = 3\,nm$) of (a) a round-robin sample of polycyclic aromatic hydrocarbons and (b) a synthetic mixture containing the same compounds. Peak assignments: 1, phenol; 2, biphenyl; 3, acenaphthene; 4, dibenzothiophene; 5 and 6, photodecomposition products of chrysene; 7, chrysene.

demonstrated by the similarity between the spectra of the reference sample and a mixture having the measured composition given in Table II.

3. Applications to Phosphorimetry

It has recently been demonstrated that the synchronous excitation concept can be applied profitably to phosphorescence, as well as fluorescence, measurements.[9] Of special interest is the application of SL to

TABLE II. Quantitative Analysis of a Complex Sample by Second-Derivative Synchronous Luminescence

Compound	d^2SL peak (nm)	Measured concentration (mg/20 ml)	Actual concentration (mg/20 ml)
Acenaphthene	328	13.7 ± 0.4	14
Biphenyl	308	15 ± 3	13.6
Chrysene	368	16.1 ± 1	15.7
Dibenzothiophene	335	12.4 ± 0.6	12.7
Phenol	285	12.0 ± 1	16.7

room-temperature phosphorimetry (RTP). Room-temperature phosphorimetry is a relatively new analytical technique based on phosphorescence at room temperature from organic compounds adsorbed on filter paper.[30] The RTP method, as opposed to conventional low-temperature phosphorimetry, does not require the use of cryogenic equipment. It allows the direct characterization of compounds collected on various substrates. Solutions from environmental samples can also be directly spotted on filter paper and the solvent evaporated under an infrared lamp prior to measuring phosphorescence.

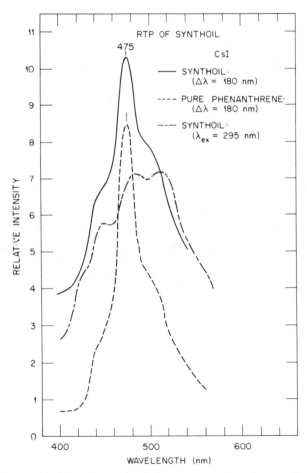

FIGURE 15. Comparison of synchronously excited RTP spectra of a coal liquid (Synthoil) and a constituent of that coal liquid sample (phenanthrene). Reprinted by permission from Vo-Dinh *et al.*[25] Copyright by Elsevier-Scientific Publishing Co.

In phosphorimetry, use can be made of the singlet–triplet splitting (Δ_{ST}) properties of organic compounds. Our research has revealed that this parameter, Δ_{ST}, can be effectively exploited as an additional factor of selectivity.[4,9] Selection of optimal values for $\Delta\lambda$ (to match the parameter Δ_{ST}) should enhance the corresponding synchronous phosphorescence signal, the spectral overlap from other components in the mixture being decreased. The physical parameter Δ_{ST} depends upon the size and chemical nature of the physical compounds. By exploiting the discrete singlet–triplet separation of each compound, this novel approach offers another possibility to analyze specific components in mixtures selectively, which should be useful in those situations where conventional fixed-excitation methods are ineffective.

An example of application of the SL technique to the RTP analysis of Synthoil is the identification of some polycyclic aromatic compounds that cannot be easily characterized by the conventional fixed-excitation method. Figure 15 shows the fixed-excitation and synchronous phosphorescence spectra of synthoil.[31] The values $\Delta\lambda = 180$ nm and $\lambda_{ex} = 295$ nm are optimal in value for phenanthrene. Whereas the fixed-excitation RTP spectrum shows a broad band at 475 nm with significant spectral overlap from the neighboring emissions from fluorene and chrysene, the synchronous spectrum exhibits a much sharper peak, thereby allowing a more accurate identification and quantification of phenanthrene.

The ability of SL and RTP to identify and quantify selected PNA compounds in a raw coal liquid (solvent refined coal) has been demonstrated in an analysis that does not require any prefractionations or separation procedures.[32] The analysis was part of a round-robin analysis organized jointly by the National Bureau of Standards (NBS) and the Department of Energy (DOE) to evaluate the efficacy of various analytical methods employed by different laboratories. Table III gives the results of the quantitative determinations by synchronous fluorescence and RTP compared with the data provided by NBS using gas chromatography–mass spectrometry (GCMS) and/or high-pressure liquid chromotography (HPLC).[32,33]

D. CONCLUSION

The complexity of real-life samples is one of the most demanding tests and one of the most serious challenges for any analytical technique. In many cases of real-life analyses, selectivity rather than sensitivity is the major problem. When it comes to analyzing multicomponent mixtures, the SL technique offers a significantly improved approach for luminescence anal-

TABLE III. Results of the Direct Determination of PNA Compounds in a Raw SRC II Sample by Synchronous Fluorescence and Room Temperature Phosphorescence[32]

Compound	Concentration (mg/g)		
	SF	RTP	NBS Data[a]
Anthracene	1.7 ± 0.2	N[b]	NA[c]
2,3-benzofluorene	1.8 ± 0.2	1.8 ± 0.6	NA
Benzo[a]pyrene	0.13 ± 0.03	N	0.134
Benzo[e]pyrene	N	0.12 ± 0.03	0.143
Carbazole	N	4 ± 1	1.96
Dibenzothiophene	N	1.0 ± 0.3	1.02
Fluoranthene	5 ± 1.5	5 ± 1.5	3.30
Fluorene	1.5 ± 0.3	N	NA
Perylene	0.02 ± 0.006	N	0.026
Pyrene	6 ± 2	3.6 ± 0.6	6.0

[a] The NBS data are of the preliminary nature (LC and/or GC–MS). The standard deviations of the NBS results range within 5–10%.[33]
[b] N, emission not detected or not resolved.
[c] NA, data not available.

ysis. The main strong point of the technique is an improved specificity that is not provided at the expense of simplicity. Despite some limitations caused by interferences inherent to luminescence spectroscopy, numerous studies have demonstrated that the SL technique is an efficient and useful analytical tool for many applications. The method has an especially great potential in analyses where simplicity, rapidity, and cost effectiveness are sought. With the renewed interest in selective and practical analytical techniques that can be used on a large-scale basis, the full impact of this method has yet to be realized.

ACKNOWLEDGMENT

This work is sponsored by the Office of Health and Environmental Research, U.S. Department of Energy under contract W-7405-eng-26 with the Union Carbide Corporation.

REFERENCES

1. C. A. Parker, *Photoluminescence of Solutions* (Elsevier, Amsterdam, 1968).
2. J. B. F. Lloyd, *Nature* (*London*) **231**, 64 (1971).
3. T. Vo-Dinh, *Anal. Chem.* **50**, 396 (1978).

4. T. Vo-Dinh, R. B. Gammage, A. R. Hawthorne, and J. H. Thorngate, *Environ. Sci. Technol.* **12**, 1297 (1978).
5. J. B. F. Lloyd and I. W. Evett, *Anal. Chem.* **49**, 1711 (1977).
6. J. B. F. Lloyd, *Analyst* **100**, 82 (1975).
7. P. John and I. Soutar, *Anal. Chem.* **48**, 520 (1976).
8. T. Vo-Dinh, R. B. Gammage, and A. R. Hawthorne, *Polynuclear Aromatic Hydrocarbons*, P. W. Jones and P. Leber, eds. (Ann Arbor Scientific Publishing Co., Ann Arbor, Michigan, 1979), p. 111.
9. T. Vo-Dinh and R. B. Gammage, *Anal. Chem.* **50**, 2054 (1978).
10. T. Vo-Dinh, R. B. Gammage, A. R. Hawthorne, and J. H. Thorngate, Proceedings of the 9th Materials Research Symposium, Gaithersburg, Maryland, April 10–13, 1978.
11. I. B. Berlman, *Handbook of Fluorescence Spectra of Aromatic Molecules* (Academic Press, New York, 1965).
12. H. W. Latz, A. H. Ullman, and J. D. Winefordner, *Anal. Chem.* **50**, 2148 (1978).
13. J. B. F. Lloyd, *Anal. Chem.* **52**, 189 (1979).
14. G. F. Kirkbright and C. G. De Lima, *Analyst* **101**, 566 (1976).
15. E. L. Wehry and G. Mamantov, *Anal. Chem.* **51**, 643A (1979).
16. J. B. F. Lloyd, *Analyst* **102**, 782 (1977).
17. T. Vo-Dinh, "Synchronous Luminescence Spectroscopy—Recent Applications and Analytical Potential," Proceedings of the 6th Annual FACSS Meeting, Philadelphia, Pennsylvania, 1979.
18. E. R. Weiner, *Anal. Chem.* **50**, 1584 (1978).
19. J. H. Rho and J. L. Stuart, *Anal. Chem.* **50**, 620 (1978).
20. Y. Talmi, D. C. Baker, J. R. Jadamec, and W. A. Saner, *Anal. Chem* **50**, 950A (1978).
21. J. B. F. Lloyd, *J. Forensic Sci.* **11**, 83 (1971); *ibid.* **11**, 153 (1971); *ibid.* **11**, 235 (1971).
22. L. P. Giering and A. W. Hornig, *Am. Lab.* **9** (11), 113 (1977).
23. T. Vo-Dinh, R. B. Gammage, and A. R. Hawthorne, *Polynuclear Aromatic Hydrocarbons*, P. W. Jones and P. Leber, eds. (Ann Arbor Scientific Publishing, Co., Ann Arbor, Michigan, 1979).
24. U.S. Environmental Protection Agency, "Environmental Assessment Sampling and Analysis: Phased Approach and Techniques for Level I," EPA-600/2-77-115 (1977).
25. T. Vo-Dinh, R. B. Gammage, and P. R. Martinez, *Anal. Chem.* **55**, 253 (1981).
26. J. C. André, M. Bouchy, and M. Niclouse, *Anal. Chim. Acta* **92**, 369 (1977).
27. D. G. Gillespie, Imperial Tobacco Ltd., Bristol, United Kingdom, private communication.
28. R. O. Rahn, S. S. Chang, J. M. Holland, T. J. Stephan, and L. H. Smith, *J. Biochem. Biophys. Methods* **3**, 285 (1980).
29. T. C. O'Haver, *Anal. Chem.* **51**, 91A (1979).
30. T. Vo-Dinh and J. D. Winefordner, *Appl. Spectrosc. Rev.* **13**, 261 (1977).
31. T. Vo-Dinh, R. B. Gammage, and P. R. Martinez, *Anal. Chim. Acta* **118**, 313 (1980).
32. T. Vo-Dinh and P. R. Martinez, *Anal. Chim. Acta* **125**, 13 (1981).
33. H. S. Hertz, National Bureau of Standards, Washington, D.C., private communication.

Chapter 6

Low-Temperature Fluorometric Techniques and Their Application to Analytical Chemistry

E. L. Wehry and Gleb Mamantov

A. INTRODUCTION

The use of cryogenic samples in analytical molecular photolumines-cence spectrometry is, in one sense, a well-established and virtually routine procedure, inasmuch as the necessity of using rigid media to detect phos-phorescence from most molecules has long been recognized.[1] Hence, glassy frozen solutions, usually at 77°K, have (until the recent advent of the solid-surface room-temperature phosphorescence procedure[2]) been regarded as de rigueur for the measurement of phosphorescence.[3] More recently, it has become evident that the use of low-temperature techniques can also offer important advantages for the measurement of fluorescence in analytical systems. It is the purpose of this chapter to examine the various techniques used, or that could be used, in analytical low-temperature fluorometry, and to consider the advantages and difficulties that characterize each of these procedures. Some moderately speculative comments regard-ing future trends in this general area are also offered.

B. GENERAL ANALYTICAL ADVANTAGES OF LOW-TEMPERATURE FLUOROMETRY

There are two fundamental reasons why one may find it desirable to em-ploy low-temperature techniques in analytical fluorescence spectrometry.

E. L. WEHRY and GLEB MAMANTOV • Department of Chemistry, University of Tennessee, Knoxville, Tennessee 37916

First, improved spectral resolution is often achieved by use of low-temperature techniques. Electronic spectra of large molecules in liquid solutions or the gas phase (except at low pressures) are often broad and virtually structureless. Such spectra are of limited value for qualitative analysis. Moreover, in samples containing several fluorophores, the fluorescence spectra of the various constituents are likely to overlap, complicating the process of quantitation. Because the absorption spectra of these compounds will also be broad and featureless, the possibilities for selective excitation of fluorescence from individual compounds in multi-component samples—which is commonly touted as a major advantage of fluorometric over absorption techniques—are actually rather limited in solution fluorometry. Any technique that increases resolution both in absorption and fluorescence spectra should enhance the selectivity of fluorometric analysis.

Second, the quantitative reliability of fluorometric analysis in mixtures is often vitiated by inner-filter effects[4] (absorption of fluorescence emitted by one compound, either by other molecules of that same compound or by those of another sample constituent) or by fluorescence quenching. Fluorescence quenching can proceed by several different mechanisms.[5] In long-range (Förster) energy transfer, excitation energy is transferred from donor to acceptor by a dipole–dipole interaction over distances that are large compared with molecular diameters. Theory predicts that this process will occur with appreciable efficiency only if the absorption spectrum of the acceptor overlaps the fluorescence spectrum of the donor. In contrast, various short-range quenching processes (which may or may not involve actual transfer of excitation energy to the quencher from the quenchee) require diffusive encounter of the interacting species.[6] Inner-filter effects can be reduced in severity by using solvent media in which highly structured absorption and emission spectra are obtained; the efficiency of long-range energy transfer is also reduced in media that provide high spectral resolution. Collisional quenching processes can be suppressed by increasing the viscosity of the solvent so that the rate constant for fluorescence exceeds that for diffusive encounter. This latter argument suggests in an obvious way that solid media (not necessarily low-temperature solids) should be used in fluorometric analysis of multicomponent samples whenever diffusional fluorescence quenching is known or suspected to be a problem.

At this point, it is necessary to examine somewhat more closely the matter of spectral resolution and bandwidths in the electronic spectra of large molecules as a function of their environment. The width of a particular band in the spectrum of an analyte molecule in a specified matrix can be thought of as consisting of two contributions.[7] First, any spectral transition has a certain inherent energy uncertainty, ultimately related in the limiting case to the excited-state lifetime through constraints imposed by the Heisenberg uncertainty principle. This "inherent" contribution to the observed spectral band width, termed *homogeneous broadening*, depends in a rather complex way upon the energy-level structure and excited-state decay properties (both radiative and nonradiative) of the analyte molecule as well as the properties of the matrix in which the analyte is situated.[8,9] In any conventional molecular fluorescence measurement, the homogeneous bandwidth defines the limit of spectral resolution that can be achieved, assuming that all molecules of the compound in question interact in exactly the same way and with precisely the same energy with their respective environments. This last condition requires in essence that each analyte molecule experience the same local environment on the fluorescence time scale. In any medium other than an absolutely perfect and rigid crystal, this situation is not likely to occur. In general, different molecules of an analyte, experiencing slightly (or perhaps drastically) different local microenvironments within the sample, will exhibit different energies of interaction with the matrix material. Consequently, the electronic transition energies for different molecules of the same compound will be somewhat different, and the resulting spectrum of the sample (which can be thought of as a weighted average of the individual spectra of the molecules comprising the sample) will be broadened by an additional amount. Spectral broadening produced by environmental heterogeneity at the molecular level is termed *inhomogeneous broadening*. In most luminescence measurements of interest to analytical chemists, the extent of inhomogeneous spectral broadening is huge compared with that of homogeneous broadening. Thus, the most intelligent analytical uses of low-temperature techniques in analytical luminescence spectrometry will tend to be directed, at least in part, toward minimization of inhomogeneous broadening.

One can visualize three rather obvious general approaches to decreasing the extent of inhomogeneous broadening in the fluorescence spectrum of a particular analyte. First, the smaller the absolute interaction energy of the analyte molecule with its environment, the smaller will be the absolute "spread" of these interaction energies summed over all analyte molecules. Therefore, it frequently pays to consider use of relatively "inert" matrix materials with which the analyte is unlikely to exhibit highly energetic interactions. For example, in the spectroscopy of phenols, the use of very polar hydrogen-bonding matrices would not be desirable. Second, the extent of inhomogeneous broadening in molecular spectroscopy can be reduced by decreasing the extent of microenvironmental heterogeneity— i.e., by making the matrix behave in as similar a manner as possible to a perfect crystal. Finally, one may be able to accept the existence of appreciable inhomogeneous broadening in the absorption spectrum of an ensemble of analyte molecules, without necessarily having that condition fully reflected in the fluorescence spectrum, by employing an excitation source having a linewidth that is substantially smaller than the width of the inhomogeneously broadened absorption spectral band. Under certain circumstances, this procedure produces site-selective excitation,[7] wherein only those analyte molecules having very similar microenvironments can be excited. The extent to which the emission spectrum suffers from inhomogeneous broadening may thus be decreased. In the following sections techniques that embody one or more of these approaches to improved spectral resolution under analytically realistic conditions will be emphasized.

C. THE SHPOL'SKII EFFECT

1. Systematics of the Phenomenon

It was first reported by Shpol'skii and co-workers in the early 1950s that highly resolved ("quasilinear") fluorescence spectra of polycyclic aromatic

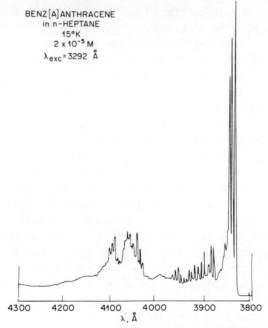

BENZ[A]ANTHRACENE
in n-HEPTANE
15°K
2×10^{-5} M
$\lambda_{exc} = 3292$ Å

4300   4200   4100   4000   3900   3800
λ, Å

FIGURE 1. Fluorescence spectrum of benz[a]anthracene in n-heptane frozen solution at 15°K.

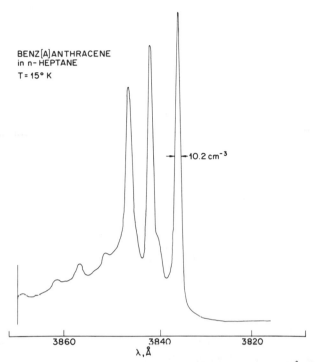

FIGURE 2. Expanded view of the triplet of "quasilines" at ca. 3840 Å in Figure 1.

hydrocarbons can be observed, provided that certain specific solvents (usually *n*-alkanes) were used and the spectra were measured at 77°K or lower temperatures.[10] Since those early observations, this phenomenon (which has come to be known as the Shpol'skii effect) has received extensive study in terms of both its fundamental chemical physics and its application to chemical analysis. An example Shpol'skii fluorescence spectrum, that of the polycylic aromatic hydrocarbon benz[*a*]anthracene in frozen *n*-heptane solution at 15°K, is shown in Figures 1 and 2; the fluorescence spectrum of the same solute in the same solvent at room temperature is shown in Figure 3. Obviously, the temperature decrease from 300 to 15°K effects an enormous increase in spectral resolution.

The general characteristics of a typical Shpol'skii fluorescence spectrum, most of which are exemplified by the spectrum shown in Figure 1, include the following:

(a) The lowest-wavelength portion of the spectrum consists of one very narrow band, or a series of narrow, closely spaced multiplets (e.g., the triplet at ~3840 Å for benz[*a*]anthracene shown in Figures 1 and 2). These

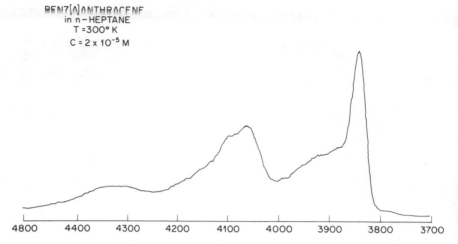

FIGURE 3. Fluorescence spectrum of benz[*a*]anthracene in *n*-heptane liquid solution at 300°K.

quasilines usually exhibit bandwidths [FWHM (full width at half-maximum)] of 15 cm^{-1} or less.

(b) At longer wavelengths a series of additional features (usually superimposed on a broad, rather diffuse band) is observed.

(c) The identity of the solvent is rather critical in the observation of truly quasilinear fluorescence spectra.[11] For example, Figure 4 shows fluorescence spectra of pyrene at 77°K in six *n*-alkanes, ranging from C$_5$ to C$_{10}$.[12] The precise appearance of the spectrum and the spectral resolution observed obviously depend strongly on the solvent. The most impressive spectral resolution is achieved when there is a rather close match between the length of the alkane carbon skeleton and the longest dimension of the aromatic solute molecule.[11,13]† Shpol'skii effects can be observed in solvents other than straight-chain alkanes; for example, quasilinear fluorescence spectra of polycyclic aromatic hydrocarbons (PAHs) and several nitrogen and oxygen heterocycles have been observed at 77°K in tetrahydrofuran.[15]

(d) The spectral resolution achieved in Shpol'skii matrices tends to improve as the sample temperature is reduced. Although the vast majority of reported quasilinear fluorescence spectra have been measured at 77°K, studies of the type shown in Figure 5 indicate the advantages to be gained in spectral resolution by use of lower temperatures.[16]

---

† However, the validity of this so-called "key and hole" rule has been questioned by Dekkers *et al.*[14]

The following picture of the origin of the Shpol'skii effect is generally accepted. A solid n-alkane matrix prepared by rapid freezing of the corresponding liquid is polycrystalline rather than glassy. In fact, single-crystal n-alkane solids containing aromatic solutes can be prepared by very slow cooling of the corresponding liquid solution.[17] The limited extent of inhomogeneous broadening of the quasilines indicates that at least some of the solute molecules occupy substitutional sites within the alkane polycrystalline lattice.[18] Hence in Shpol'skii spectra the extent of inhomogeneous broadening is decreased by incorporating the various molecules of the fluorophore in very similar microenvironments.

Several lines of evidence are consistent with this picture. In the first place, substituted or nonplanar aromatic solutes usually do not yield spectra of high resolution in Shpol'skii matrices.[19,20] Moreover, Lamotte and Joussot-Dubien have noted that both the absorption and fluorescence spectra of polycylic aromatic hydrocarbons incorporated into n-alkane single crystals are strongly polarized, indicating a high degree of orientational specificity of the fluorophore molecules in the solid alkane. Analysis of the polarized absorption and fluorescence spectra permits the preferred orientation of the long axis of an aromatic hydrocarbon solute to be estimated.[17] More precise numerical estimates of this orientational specificity in alkane single crystals have been obtained by analysis of ESR spectra[21,22] and polarized microwave-optical double resonance measurements[23] of solute triplet states. It is generally believed that planar aromatic compounds are oriented in the alkane lattice with their aromatic planes parallel to the unfolded carbon skeleton chains of the alkane molecules[17]; an example representation of one such situation (coronene in crystalline n-heptane) is shown in Figure 6. The coronene molecule is believed to occupy a site created by the displacement of three heptane molecules; smaller aromatic solutes can be accomodated substitutionally in a site formed by replacement of two heptane molecules.[24]

Two additional characteristics of Shpol'skii fluorescence spectra have attracted considerable attention: the multiplet structure of the quasilines (present in most but not all Shpol'skii spectra) and the relatively broad emission at longer wavelengths (e.g., at ~4050 Å in the benz[a]anthracene spectrum shown in Figure 1). When present, the multiplet structure is seen in excitation as well as emission spectra. At sufficiently low temperatures and using narrow-bandwidth laser light, it is possible to selectively excite specific sets of multiplets, as shown by the example in Figure 7. This result indicates that the multiplets correspond to sets of molecules contained in specific, different types of sites within the solid matrix.[25–28] It was initially believed that the multiplet structure indicated the existence of different polymorphic domains within the quasicrystalline matrix,[29] but it was subsequently

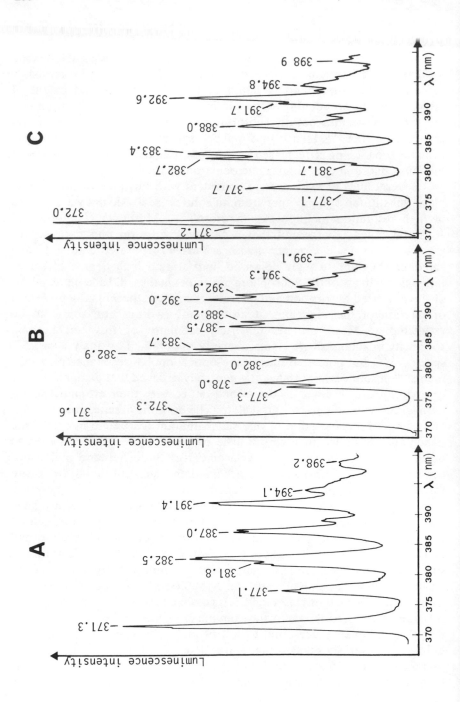

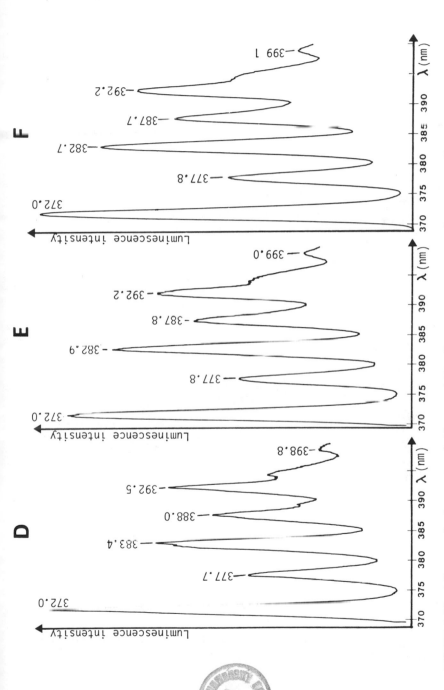

FIGURE 4. The influence of solvent upon the frozen-solution fluorescence spectrum of pyrene at 77°K. All n-alkanes: (A) pentane, (B) hexane, (C) heptane, (D) octane, (E) nonane, (F) decane. From Colmsjö and Stenberg[12]; reproduced by permission of Almquist and Wiksell.

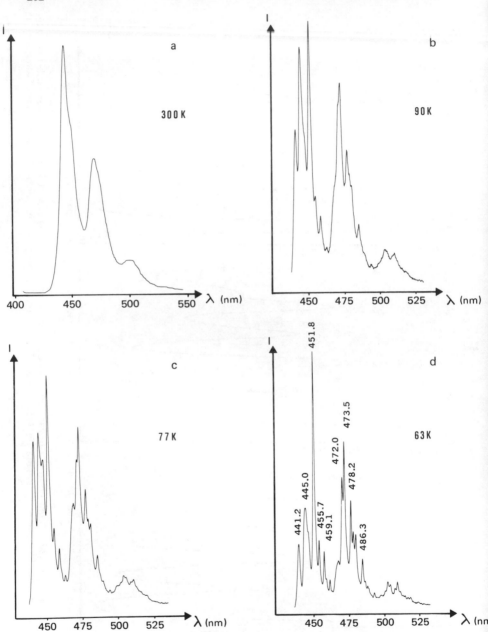

FIGURE 5. Effect of temperature upon fluorescence spectrum of perylene in *n*-heptane frozen solution. From Colmsjö and Stenberg[16]; reproduced by permission of Almquist and Wiksell.

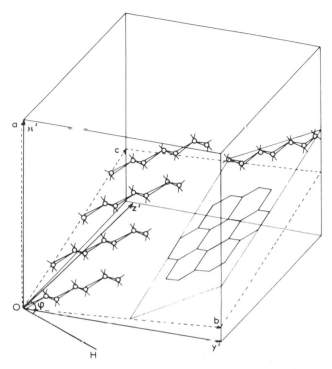

FIGURE 6. Representation of the orientation of a coronene molecule in a single-crystal *n*-heptane matrix. Note the orientation of the heptane chains and the replacement of three heptane molecules by one solute (coronene) molecule. From Merle *et al.*[21]; reproduced by permission of North-Holland.

demonstrated that, even in alkanes that exhibit a single stable crystallographic modification from 77°K to their melting points, multiplet structure in the quasilinear emission spectra of aromatic solutes is observed.[18]

Various other hypotheses regarding the nature of the "multiple sites" have subsequently been proposed. It is, for example, plausible that the multiple sites may correspond to several different rotational conformers of the flexible alkane chains[30,31]; evidence from matrix isolation fluorescence spectra obtained in vapor-deposited alkane media (see Sec. D.2) is consistent with that viewpoint. Alternatively, there may be several well-defined but slightly different substitutional orientations of molecules of a particular PAH within an essentially regular alkane lattice.[32] There is evidence that solute molecules having different substitutional positions in the matrix suffer different types and degrees of geometric distortion in ground and/or excited states, with these distortions being reflected as site splittings in the

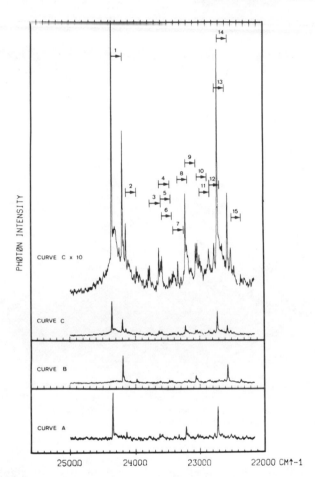

FIGURE 7. Phosphorescence spectra of N-ethylcarbazole (10^{-4} M) in n-heptane at 4°K: Curve A, excitation at 28,980 cm^{-1} (bandwidth; 30 cm^{-1}); curve B, excitation at 28,830 cm^{-1} (bandwidth; 30 cm^{-1}); curve C, excitation at 28,900 cm^{-1} (bandwidth; 300 cm^{-1}). The excitation source was a 2500-W xenon lamp dispersed by a 0.75-m double monochromator. From Vo-Dinh, Kreibich, and Wild[26]; reproduced by permission of North-Holland.

spectra.[24,28,33,34] Indeed, for the specific case of the aromatic hydrocarbon coronene in n-heptane Shpol'skii matrices it has been contended that "the spectral properties ... are determined primarily by the static distortions of the molecule due to the presence of different crystal fields."[28] This question is far from being conclusively settled; in the words of Clar *et al.*, "A great number of experiments has been performed in order to better understand the conditions leading to such site effects, without giving final results."[34]

The origin of the diffuse bands in Shpol'skii fluorescence spectra is now rather well understood. The "quasilinear" portions of a Shpol'skii fluorescence spectrum correspond to purely electronic transitions of the solute— that is, to transitions in which the vibrational state of the lattice does not change. Quanta of lattice vibrational energy are commonly termed "phonons." Hence, the quasilinear portion of a Shpol'skii fluorescence spectrum (such as the triplet at ~ 3840 Å in Figure 1) is referred to as the "zero phonon region" of the spectrum. On the long-wavelength side of a zero-phonon fluorescence band (and on the short-wavelength side of the corresponding zero-phonon region in the absorption spectrum) occurs a "phonon wing," wherein the electronic transition in the solute is accompanied by changes in the vibrational state of the lattice.[35-38] Indeed, Rebane[37] has suggested an analogy between the origin of the zero-phonon quasilines in an optical spectrum and the Mössbauer effect (i.e., recoil-less emission and resonant absorption of γ rays by certain nuclei in solid samples).

One effect of the solvent lattice, therefore, is to provide a large number of additional vibrational degrees of freedom coupled to the electronic transitions of the solute species. It is desired that the degree of such coupling of the solute and lattice be weak[36]; otherwise, the zero-phonon lines will be weak and difficult or impossible to detect. The intensity ratio of the zero-phonon line to that of the phonon sideband is commonly known as the "Debye–Waller factor."[37] This quantity decreases rapidly with increasing temperature[39]; accordingly, the observation of an intense zero-phonon band in the presence of a weak phonon wing is often possible only at very low temperatures (frequently 20°K or less). Fortunately, the degree of coupling of polycyclic aromatic hydrocarbons to lattice vibrational modes in frozen polycrystalline n-alkanes is usually quite small[36]; thus, an intense zero-phonon line is usually observed at 77°K in Shpol'skii solvents. For other solutes, such as porphyrins[40], lower temperatures may be required in order to observe the characteristic quasilinear character of Shpol'skii fluorescence spectra.

2. Analytical Applications

The possibility that Shpol'skii fluorescence spectra could serve as effective "fingerprints" for identification of specific fluorophores became evident almost immediately upon discovery of the phenomenon, and numerous applications of the procedure to qualitative analyses were reported by Soviet workers in the 1960s; this literature has been reviewed by Kirkbright and de Lima.[41] Probably the most thorough evaluation of conventional Shpol'skii fluorometry for analytical purposes has been carried

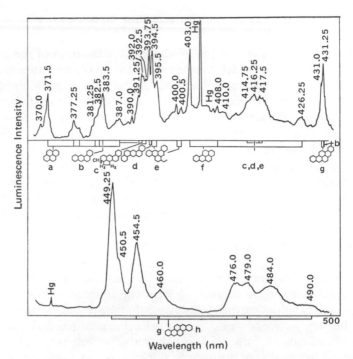

FIGURE 8. Luminescence spectrum of a seven-component synthetic mixture of poly-cyclic aromatic hydrocarbons in *n*-octane: cyclohexane frozen solution at 77°K. The band systems corresponding to the various compounds, along with the structures of the compounds, are indicated. From Kirkbright and de Lima[41]; reproduced by permission of the Chemical Society.

out by Kirkbright and co-workers.[41–44] For example, Figure 8 shows a Shpol'skii fluorescence spectrum of a synthetic mixture of eight polycyclic aromatic hydrocarbons (including two sets of isomer pairs); the spectral resolution is sufficient to permit unambiguous identification of all eight compounds, which could not possibly be accomplished by conventional solution fluorometry without prior separation. Qualitative analyses of a number of real samples (including solvent extracts of coal and coal–tar pitch,[44] extracts of organic matter from geological samples,[45] and jet airplane engine emissions[46]) by Shpol'skii fluorometry have been reported. The Shpol'skii technique has also been used to examine fractions produced by liquid chromatography of real samples such as automobile exhaust.[47] An example fluorescence spectrum of one such liquid chromatography fraction is shown in Figure 9. The use of Shpol'skii luminescence spec-trometry to examine compounds separated by thin-layer chromatography (TLC) (by vacuum-subliming the analytes from a TLC plate onto a cold finger which is then washed with a Shpol'skii solvent) has also been described.[48]

Numerous applications of fluorescence in Shpol'skii matrices to quantitative analysis, especially in synthetic mixtures, have been reported. Here, as in most other applications of molecular luminescence to quantitative analysis, some type of empirical standardization procedure is usually required. Kirkbright and de Lima[41] have concluded that the most precise quantitative results are obtained by use of a combined standard addition–internal standard technique. Provided that appropriate care is taken in the design of the cell holder, in reproducible positioning of the sample cell in the optical path of the fluorescence spectrometer, and in achieving reasonable reproducibility in the rate at which the initial liquid solution is frozen, relative standard deviations as low as 2% can be achieved in quantitative analyses by Shpol'skii fluorometry.

A significant innovation in Shpol'skii luminescence spectrometry has been introduced by Fassel, D'Silva, and their co-workers, who use x-rays rather than UV or visible light to excite emission from samples dispersed in frozen-solution matrices.[49–51] An example spectrum obtained for a complex sample by this technique is shown in Figure 10. Several advantages

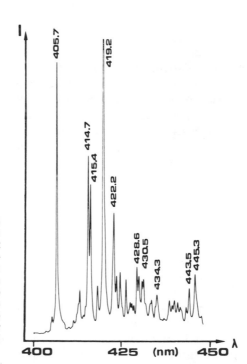

FIGURE 9. Fluorescence spectrum of a liquid chromatography fraction of an extract from an automobile exhaust sample. Solvent; *n*-hexane; temperature: 63°K; excitation wavelength; 313 nm. The positions of the main peaks in this spectrum indicate the principal component of this fraction to be benzo[*ghi*]perylene. From Colmsjö and Stenberg[47]; reproduced by permission of the American Chemical Society.

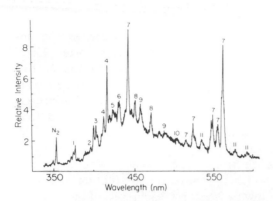

FIGURE 10. X-ray induced luminescence spectrum of extract from airborne particulate matter in *n*-heptane at 90°K. Identified compounds: (1) chrysene; (2) dibenz[*a,h*]anthracene; (3) benzo[*a*]pyrene; (4) benzo[*ghi*]perylene: (5) anthracene; (6) dibenzo[*a,i*]pyrene; (7) coronene; (8) perylene; (9) triphenylene; (10) phenanthrene; (11) benzo[*e*]pyrene. From Woo, D'Silva, and Fassel[51]; reproduced by permission of the American Chemical Society.

of x-rays over conventional optical illumination are evident. First, scattered source light is not likely to be a serious problem when x-ray excitation is used; scattering is often a significant problem in conventional Shpol'skii fluorometry because the frozen solutions are polycrystalline, rather than glassy, and are therefore efficient scatterers. Second, x-ray illumination does not produce excitation by the same mechanisms as does UV; hence, optically forbidden excited states may be populated by x-ray illumination and the luminescence properties of molecules may be quite different than those observed with conventional optical excitation. Phosphorescence tends to be a relatively more important radiative decay mode for certain analytes (e.g., polycyclic aromatic hydrocarbons) under x-ray illumination than in conventional luminescence spectroscopy. This fact is particularly relevant whenever it is desired to carry out time-resolved luminescence measurements, because temporal resolution is experimentally much more straightforward for phosphorescence than for fluorescence.[52] Finally, for very highly colored samples, such as shale oil or coal liquid extracts, x-ray illumination has obvious advantages. The disadvantages of x-ray excitation include the cost of an x-ray source (not a common item of apparatus in a conventional optical spectroscopy laboratory), the lack of literature data for luminescence properties of organic molecules (other than liquid scintillator systems) under x-ray illumination, the lack of fundamental understanding of the processes by which x-ray illumination leads to generation of electronically excited states of organic molecules, and the possibility of x-ray illumination causing the sample to undergo radiolysis, which would alter the sample composition as a function of time. Since many compounds undergo photolysis in Shpol'skii matrices under UV illumination, this latter problem is obviously not restricted to x-ray induced luminescence. Probably the most severe shortcoming of x-ray illumination is that there is no obvious way in which individual fluorophores in complex mixtures can be selectively excited. Hence, the x-ray technique may be more useful for qualitative analyses (particularly for rapid surveys of the polycyclic aromatic content of very complex samples) than for quantitation of specific compounds in mixtures. Further developments in the analytical use of x-ray induced Shpol'skii luminescence are anticipated.

While Shpol'skii fluorescence spectrometry has interesting analytical potential (not all of which has yet been effectively exploited), it is not devoid of shortcomings, many of which have been itemized by Lukasiewicz and Winefordner.[53] As noted above, the Shpol'skii effect is a solvent-specific phenomenon. In a complex sample, not all components will exhibit Shpol'skii-quality spectra in any one solvent. In practice, this is not likely to constitute a significant problem, since for any sample of extreme complexity the time required to obtain fluorescence spectra in several solvents is not likely to be a major contributor to the total analysis time (including sample cleanup, separations, etc.). In fact, the time saved by the fact that multi-component samples can be dealt with in Shpol'skii matrices (owing to the high spectral resolution), and therefore less efficient separations can be used, may be appreciable even if several different solvents must be used in sequence to extract all the required spectroscopic data.

More serious are the facts that both intensities and bandwidths in Shpol'skii spectra are in some cases dependent upon the freezing rate[54–58] or the final temperature[55] of the frozen solution. Depending upon the rate at which freezing is effected, Shpol'skii solid solutions may not represent "equilibrium" systems; the closer the matrix is to true thermodynamic equilibrium, the greater are the spectral bandwidths typically observed in Shpol'skii media.[53] Bandwidths in Shpol'skii spectra are in some cases dependent upon the analyte concentration,[13,55,58,59] and the relative intensity of the zero-phonon line to phonon sideband regions is often strongly temperature dependent.[58,60] The appearance of the Shpol'skii fluorescence spectrum of a particular compound may depend upon the identities and concentrations of other sample constituents.[54,61]

In most of the cases in which analytical calibration curves have been reported for quantitative Shpol'skii fluorometry, the linear region is rather restricted. There appears to be only one report[62] in which a linear dynamic range exceeding three decades in analyte concentration has been claimed for Shpol'skii fluorescence spectrometry; other quantitative studies[41,42,63–65] have reported a linear working range of 2.5 decades or less. It is not clear that this problem is really as serious as the preceding sentences imply, or whether studies attempting to establish a broad linear dynamic range for Shpol'skii fluorometry have simply not been performed or have been executed with insufficient care. There are clear indications that inner-filter effects and intermolecular electronic energy transfer can occur in Shpol'skii matrices containing mixtures of fluorescent solutes,[11] and these matters must be explored more thoroughly before the quantitative potentialities of the Shpol'skii effect can be properly assessed.

A general problem for any frozen-solution medium (including the common glassy media used for phosphorescence, as well as polycrystalline Shpol'skii matrices) is the fact that solubilities of most compounds in liquid solvents decrease with decreasing temperature. Hence, in formation of a Shpol'skii matrix, aggregation of solute molecules may occur unless the solute concentration is low and the freezing rate is rapid.[58,66-68] The aggregation process produces spectral broadening; in particularly severe cases, microcrystalline solute domains may be formed within the matrix. Fluorescence quenching is often very efficient within such microcrystals; other undesirable effects, such as excimer fluorescence, may also be observed.[67] The production of microcrystals in frozen solution matrices has been reported at solute concentrations as small as 10^{-5} M.[66,67] Hence, it is obviously desirable to be certain that samples (particularly complex mixtures) are quite dilute before attempting to perform quantitative analyses by Shpol'skii fluorometry. All of the effects discussed above can contribute to problems in achieving suitable linear dynamic ranges and reproducibilities in quantitative Shpol'skii fluorescence spectrometry, and further investigations of these effects are needed.

Lukasiewicz and Winefordner[53] also note that the high spectral resolution achieved in Shpol'skii matrices is beyond the capabilities of the vast majority of commercial fluorescence spectrometers. Thus, this (or any other) technique producing quasilinear spectra is unlikely to achieve "routine" analytical status in the near future. While this comment is certainly valid, the advantages of high spectral resolution in the analysis of complex samples are becoming increasingly widely recognized, and commercial fluorescence instrumentation that exhibits sufficient monochromator resolution to exploit the Shpol'skii effect is now available. Consequently, the "instrumentation" problem does not appear now as a formidable barrier to use of the Shpol'skii effect in practical analytical chemistry.

As a final comment on the Shpol'skii effect, we may note that (as discussed by Kirkbright and de Lima in 1974[41]) the phenomenon should be ideally suited to laser excitation. In liquid solutions, the absorption spectra of most organic molecules are so broad that laser excitation of fluorescence achieves little, if anything, in the way of increased selectivity (although, as discussed in Chapter 1 of this volume, spectacularly low detection limits can be achieved in favorable cases). However, in Shpol'skii matrices absorption spectra are sufficiently narrow so that selective excitation of fluorescence from individual compounds in multicomponent samples should be achieved readily by use of a tunable laser as a source. Rather surprisingly, as of this writing (April 1980) no published application of selective laser excitation or time-resolved fluorometry to complex mixture analyses by Shpol'skii fluorometry has yet appeared. However, a paper presented by Fassel and co-workers at the 1980 Pittsburgh Conference,[69] dealing with the use of dye laser excitation in Shpol'skii fluorometry at very low temperatures (15°K) for the selective excitation of fluorescence from individual polycyclic aromatic hydrocarbons in coal liquids and shale oils, demonstrates the advantages to be derived from the use of tunable laser excitation in low-temperature fluorometry. In this study, it was shown possible to excite fluorescence from several sample constituents and to quantitate these specific compounds in very complex samples without prior chromatographic separations. Indeed, it is evident that the full potential of the Shpol'skii effect in analytical chemistry will not be realized until laser excitation of fluorescence is applied as a matter of routine.

D. MATRIX ISOLATION

1. Fundamentals and Experimental Techniques

Matrix isolation spectroscopy, first described in the mid-1950s by Whittle, Dows, and Pimentel[70] and Norman and Porter,[71] has become a "standard" procedure for observation of optical and magnetic resonance spectra of transient species (including a variety of free radicals and, more recently, metal atoms and clusters). However, it is only recently that the utility of matrix isolation as a procedure for qualitative and quantitative spectroscopic analyses of stable compounds in multicomponent samples has been appreciated, largely as a result of work performed in this laboratory.[72]

The principal motivation for our studies of analytical matrix isolation spectroscopy stemmed from consideration of the properties of an "ideal" sampling medium for optical spectroscopy, and specifically from the effects enumerated in the previous section that can cause quantitative applications of the Shpol'skii effect to encounter difficulty. An "ideal solvent" for molecular fluorometry should satisfy several rather obvious criteria. First, it should be transparent to the exciting light and the fluorescence of the analyte, and it should not exhibit highly efficient Rayleigh or Raman scattering. Second, it should contain low levels of fluorescent contaminants; alternatively, it should be a fairly straightforward matter to remove luminescent contaminants if they are initially present. (This requirement arises from the fact that when modern fluorescence instrumentation is employed, and particularly if laser excitation is utilized, the factor that usually determines detection limits is the fluorescence blank of the solvent.[73] Third, as noted in the preceding discussion of the Shpol'skii effect, it is highly desirable that a sampling medium for fluorescence spectrometry produce highly structured spectra exhibiting narrow features suitable for "fingerprinting" and quantitative analysis in mixtures. The fluorescence excitation spectra should also be highly resolved in order that selective excitation of fluorescence from individual compounds in multi-component samples be feasible. The positions of the various spectral features exhibited by any one compound, and the widths and relative intensities of those features, should be unaffected (within reasonable limits) by variations in sample composition and pretreatment procedures. Finally, in order to achieve precise quantitation and wide linear dynamic range in complex samples the fluorescence quantum efficiency exhibited by a particular compound should be independent of sample composition (hence quenching must be absent).

This last requirement is probably the most stringent, because inter-molecular electronic energy transfer can occur by both collisional and long-range processes.[74] Thus, a high-viscosity "solvent" must be used (to suppress collisions). Moreover, the average distance between fluorophore molecules must be large in comparison with the "critial transfer distance" for Förster resonant energy transfer; to achieve this condition the concentration of solute molecules in the spectroscopic matrix must be small and they must be distributed within that matrix in an essentially random manner. Aggregation effects must be negligible, and the solute molecules must behave as though they were isolated from each other and experience only the matrix as part of their environment. Such conditions must also be satisfied in the spectroscopic study of high reactive chemical species. Hence, all of this suggests that matrix isolation, initially developed for the study of highly reactive entities, should also be a useful approach for fluorometric analysis of complex samples of chemically stable molecules. Interestingly, the analytical potential of matrix isolation fluorometry (using laser excitation!) was discussed as early as 1969 by Shirk and Bass[75]; nonetheless, it was not until 1977 that experimental verification of the analytical utility of matrix isolation fluorescence spectrometry was reported [76]

In matrix isolation (MI), samples that are solids or liquid at room temperature are vaporized; the resulting gaseous sample is mixed with a large excess of a diluent gas (matrix gas). One often desires the matrix gas to be relatively chemically inert; thus the conventional matrices employed in MI are the rare gases (especially argon and, to a much smaller extent, xenon and krypton) and nitrogen. In molecular fluorometry, however, chemical inertness may be superseded by other factors as the principal criterion for choice of matrix material (see Sec. D.3).

Once the sample vapor has been thoroughly mixed with the matrix gas, the mixture is deposited on a solid surface at cryogenic temperatures for spectroscopic examination in the solid state. The reason for using this (rather indirect) procedure for preparation of a low-temperature solid sample is to circumvent the aggregation effects that may occur in the preparation of solid samples by freezing of liquid solutions (see Sec. C); the essentially random molecular distribution of the vapor sample should, if retained in the cryogenic solid, prevent aggregation effects and eliminate all forms of fluorescence quenching. For this condition to actually be achieved it is necessary that the mole ratio of matrix gas to solute species by very large (10^4 or greater for fluorescence spectroscopy) and that the mixing of matrix gas with solute vapor be quite thorough. The temperature of the resulting gaseous mixture must be sufficiently high so that any tendency for dimers or higher aggregates to be present at appreciable concentrations in the vapor phase is minimal. Moreover, the viscosity of the solid deposit as it forms must be sufficiently great so that diffusive aggregation of solute molecules in the solid does not occur to an appreciable extent. The latter requirement signifies that care must be exercised in the choice of a surface for deposition (high thermal conductivity is an obvious requirement) and that the deposition rate must be controlled so that the temperature of the deposit does not rise to more than a factor of about 0.4 of the melting point of the matrix.[77] Obviously, there are advantages in choosing matrix materials that melt at temperatures which far exceed the lowest operating temperature of the cryostat used in the experiment.

Experimental techniques in matrix isolation have been reviewed in detail by several authors[8,78–81]; techniques in analytical MI spectroscopy have been surveyed by the present authors.[72] Briefly, liquid or solid samples are placed in a Knudsen cell, which is an apparatus possessing an orifice (1 mm or less in diameter) at one end. The sample is heated and the vapor so formed effuses through the orifice into an evacuated cryostat head. The matrix gas is either admitted to the cryostat head by a separate line or is swept through the Knudsen cell. A schematic diagram illustrating the former approach is shown in Figure 11. In either case, a region for vapor mixing must be provided. The mixed vapor is then deposited on a cold surface. If right-angle or in-line[82] illumination geometry is desired, such that an optically transparent deposition surface is needed, sapphire is a suitable substrate; neither optical quartz nor fused silica exhibits sufficient thermal conductivity. For application of front-surface illumination, a metallic surface such as gold-plated oxygen-free high-conductivity copper is suitable.

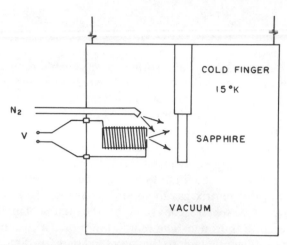

FIGURE 11. Schematic diagram of cryostat head design used for matrix isolation fluorescence spectrometry. Samples are placed in the Knudsen cell (wrapped with heating wire maintained at voltage V), mixed with matrix gas (in this case, N_2), and deposited on a cold surface (in this case, a sapphire window). The head shown is attached to a closed-cycle refrigerator. From Wehry and Mamantov[72]; reproduced by permission of the American Chemical Society.

Unless a very-high-melting matrix is used, the cold finger must be maintained at a temperature well below 77°K. Hence, the liquid nitrogen cryostats commonly used in frozen-solution fluorometry cannot ordinarily be used in MI. The technique of matrix isolation has become accessible to any laboratory, irrespective of its lack of expertise or previous experience in cryogenics, by the development of commercial closed-cycle helium refrigerators,[79] which can attain temperatures as low as 10–15°K without the use of liquid cryogens or any other apparatus more exotic than a vacuum line. As noted in Sect. C, the quality of Shpol'skii fluorescence spectra is also improved by the use of temperatures lower than 77°K, and closed-cycle refrigerators are beginning to experience use in Shpol'skii fluorometry[69] as well as in MI. For certain specialized experiments such as site selection (Sec. E), temperatures lower than 10°K are required; in such cases, use of a liquid helium cryostat is necessary. For most purposes in MI fluorometry, recourse to such low temperatures is unnecessary.

Various modifications of the Knudsen effusion procedure, used to deal with very volatile samples or very small deposition surfaces, have been described by Hembree *et al.*[83] Once the solid deposit is prepared, its fluorescence spectrum can be obtained using a fluorometer of conventional design. Three important requirements must be met by the fluorescence spectrometer. First, the sample cell area must be sufficiently spacious to

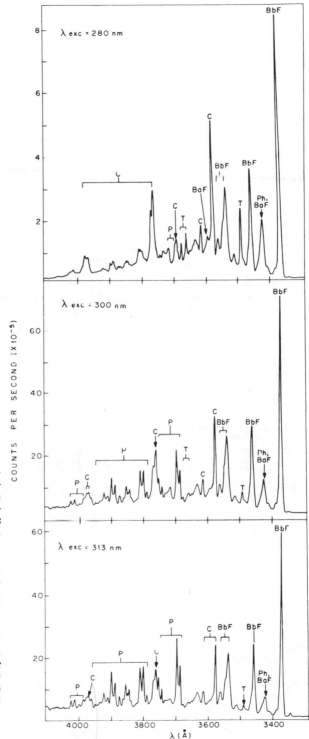

FIGURE 12. Matrix isolation fluorescence spectra at three excitation wavelengths of a six-component synthetic mixture of polycyclic aromatic hydrocarbons; matrix, N_2; temperature, 16°K. Compounds: P, pyrene (500 ng); C, chrysene (400 ng); BaF, benzo[a]fluorene (400 ng): BbF, benzo[b]fluorene (250 ng); Ph, phenanthrene (500 ng); and T, triphenylene (1.37 μg). From Stroupe *et al.*[76]; reproduced by permission from the American Chemical Society.

accommodate a cryostat; unfortunately, many commercial fluorometers are incompatible with cryostats or other relatively bulky sampling devices. Second, as emphasized for the Shpol'skii effect, the emission monochromator must have a sufficiently small spectral bandpass to avoid distortion of the quasilinear spectra that can be obtained by MI under the proper conditions. Finally, the source should exhibit a sufficiently small bandpass to be suitable for excitation of individual compounds in mixtures. For many purposes, a continuum source–monochromator combination suffices but for extremely high resolution a laser is required.

A set of example matrix-isolation spectra at three different excitation wavelengths for a synthetic mixture of polycyclic aromatic hydrocarbons is shown in Figure 12. These spectra were obtained using a continuum source (Xe–Hg lamp and excitation monochromator; bandpass = 7 nm FWHM) in nitrogen, a "conventional" matrix.[76] Note that each of the compounds can be identified without difficulty via specific, intense fluorescence bands. An example continuum-source-excited fluorescence spectrum of a real sample (an adsorption chromatography fraction from a steel mill coking plant wastewater sample), also in nitrogen, is shown in Figure 13.

One of the fundamental objectives of MI is to secure quantitative results free from quenching and (to the maximum extent possible) inner-

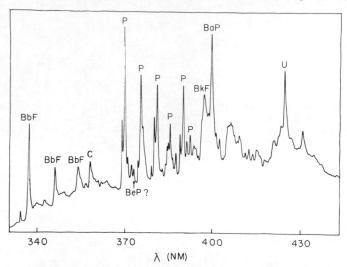

FIGURE 13. Matrix isolation fluorescence spectrum ($\lambda_{ex} = 313$ nm; Xe–Hg lamp; matrix = N_2) of absorption chromatography fraction from steel mill coking plant water sample. Identified compounds: BbF, benzo[b]fluorene; C, chrysene; P, pyrene; BeP, benzo[e]pyrene; BkF, benzo[k]fluoranthene; BaP, benzo[a]pyrene. U denotes band exhibited by unidentified compound. This sample contains at least 25 different polycyclic aromatic compounds. From Wehry and Mamantov[72]; reproduced by permission of the American Chemical Society.

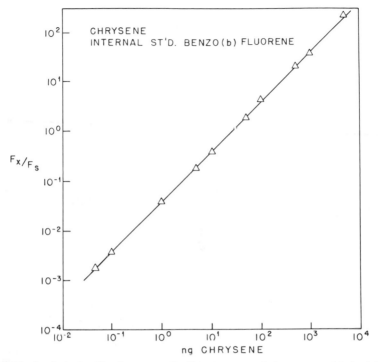

FIGURE 14. Analytical calibration curve for fluorescence of chrysene matrix isolated in N_2, using benzo[*b*]fluorene as internal standard.

filter effects. It may initially be argued that the quantitative potential of MI is rather limited, inasmuch as it is ordinarily not possible to be certain that quantitative transfer of the sample from the Knudsen cell to the deposition surface takes place. Indeed, in most applications of MI to quantitative analysis, no attempt is made to obtain quantitative transfer or to measure the extent to which the deposition is not quantitative. Instead, an internal standard (which matches reasonably closely the volatility of the analyte compounds) is used. It has repeatedly been observed that internal standardization is sufficient to ensure adequate quantitative precision and linearity; one simply assumes that the extents of "nonquantitativeness" of deposition of the analyte and standard are the same.[76,84] In most analyses of real samples, some form of internal standardization is necessary to compensate for nonquantitative sample workup steps (such as extractions); thus, an internal standardization as part of the actual spectroscopic quantitation step is generally not an unduly burdensome requirement.

Two examples indicating the linear dynamic range of MI fluorometry are shown in Figures 14 and 15. In Figure 14 is shown an analytical

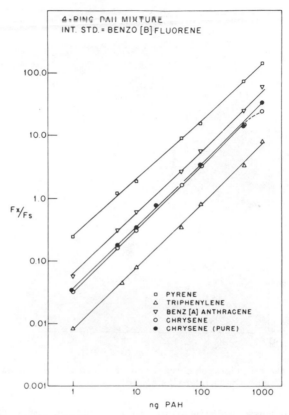

FIGURE 15. Analytical calibration curves for four-ring polycyclic aromatic hydrocarbons in four-component synthetic mixtures. For each compound the fluorescence intensity was ratioed to that of an internal standard, benzo[b]fluorene. Each sample contained equal weights of the four compounds. Shown for comparison is a calibration curve, obtained under identical conditions, for pure chrysene. From Stroupe et al.[76]; reproduced by permission of the American Chemical Society.

calibration curve for pure chrysene in nitrogen matrices, using benzo[b]fluorene as internal standard; the curve is linear from the detection limit (10 pg) to an amount of chrysene exceeding 1 μg.† The upper limit of linearity in MI fluorometry is usually on the order of 1–5 μg; its value appears to be determined by the appearance of appreciable inner-filter

---

† Note that analytical figures of merit in MI fluorescence are usually specified in quantities, rather than concentrations, of analyte (because the analyte concentration in the sample subjected to spectroscopic analysis is not known). The concentration of analyte in the original sample can, of course, be ascertained from MI spectral data if the volume or mass of the original sample introduced into the Knudsen cell is known.

absorption. The lower limit of linearity is established by the limit of detection. For cases wherein subpicogram detection limits have been observed, usually by laser excitation in unconventional matrices,[85] the linear dynamic range is greater than six decades.

The series of analytical calibration curves in Figure 15 is especially noteworthy.[76] These data were obtained from four-component synthetic samples (three of the components being isomers), in each of which equal weights of the four compounds were present. The calibration curves are linear and parallel up to ~ 1 μg of each compound. Note especially that the calibration curve for chrysene in the presence of the other three compounds is virtually superimposable upon that obtained for pure chrysene. This result provides a particularly dramatic illustration of true matrix isolation—that is, a situation wherein the fluorescence signal observed for a particular analyte is unperturbed by the presence of potentially interfering sample constituents. This example is especially striking because in liquid solution the fluorescence of chrysene is badly quenched in the presence of the other three solutes.

2. Matrix Isolation in Shpol'skii Media

Comparison of Figure 12 with any of the Shpol'skii spectra shown in Sec. C illustrates the fact that MI in conventional matrices produces spectral resolution inferior to that obtainable by continuum-source excitation in Shopl'skii frozen solutions. Typical bandwidths (FWHM) for fluorescence spectra excited by a continuum source in N_2 or argon matrices are ~ 100 cm^{-1}, whereas widths of individual quasilines in Shpol'skii spectra often are 10 cm^{-1} or less. While the resolution achieved by MI fluorometry in conventional matrices is often sufficient for analyses of multicomponent samples of moderate complexity, the analysis of very complex samples requires higher resolution in both excitation and emission spectra. This can be achieved either by use of "unconventional" matrices, such as Shpol'skii solvents, or by use of laser excitation. Each of these approaches to improved selectivity in MI fluorometric analysis will now be considered in turn.

Inasmuch as most Shpol'skii solvents are relatively volatile, there are no serious experimental obstacles to using them (or other organic solvents) as matrices in MI spectroscopy. A schematic diagram of a simple apparatus for MI in n-heptane is shown in Figure 16. If a deposit of a suitable polycyclic aromatic hydrocarbon, such as benz[a]anthracene (BaA), is formed in the conventional way by deposition of a BaA–heptane vapor mixture at 15°K, the observed spectral resolution is no greater than that observed for a BaA matrix isolated in nitrogen. However, if a deposit in a Shpol'skii matrix formed at 15°K is annealed at a higher temperature and then cooled back to

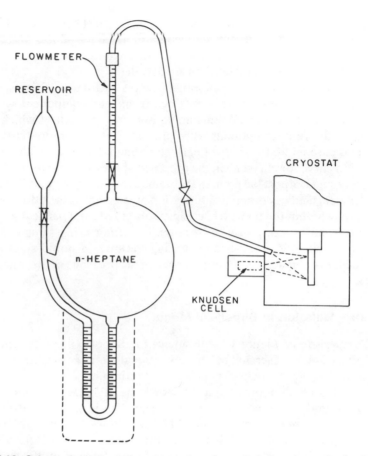

FIGURE 16. Schematic diagram for preparation of matrix-isolated samples, using the vapor of an organic solvent (such as *n*-heptane) as the matrix gas. The U tube at the bottom of the heptane reservoir permits the solvent to be degassed by freeze–thaw cycles to remove dissolved O_2 that might otherwise act as a fluorescence quencher.

15°K prior to measurement of a fluorescence spectrum, quasilinear spectra virtually identical with those observed in Shpol'skii frozen solutions are obtained.[86] These effects are illustrated in Figure 17. Of particular interest is a "memory effect" observed in the annealing process. Suppose, for example, that a deposit, initially formed at 15°K, is annealed at 140°K. If the spectrum of this deposit is measured at 140°K, broad bands are observed. Decreasing the temperature to 15°K produces quasilinear structure in the spectrum, as shown in Figure 18. If this deposit subsequently undergoes additional annealing at any temperature less than 140°K, the fluorescence

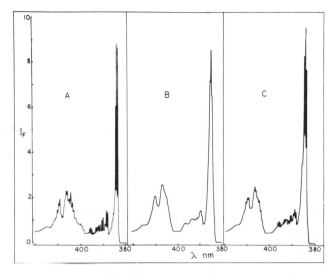

FIGURE 17. Fluorescence spectra of benz[*a*]anthracene in *n*-heptane at 16°K: (A) in frozen solution; (B) matrix isolated at 16°K, no annealing; (C) matrix isolated at 16°K, annealed at 140°K, and then returned to 16°K prior to measurement of spectra. From Tokousbalides, Wehry, and Mamantov[86]; reproduced by permission of the American Chemical Society.

spectrum observed at 15°K will continue to exhibit the original pattern of quasilines, irrespective of the number or duration of subsequent annealing steps. However, if in a subsequent annealing step a temperature higher than 140°K is used, then (upon return to 15°K) a new distribution of intensities among the zero-phonon multiplets is observed, though the positions of the

FIGURE 18. Preservation of "memory" in matrix-isolation fluorescence spectrum of benz[*a*]anthracene in *n*-heptane. Fluorescence band system at 384 nm (cf. Figure 19) recorded at (A) 16°K after annealing at 140°K; (B) sample (A) at 100°K; (C) after cooling (B) back to 16°K; (D) 16°K after annealing at 150°K; (E) sample (D) at 100°K; (F) after cooling (E) back to 16°K. From Tokousbalides, Wehry, and Mamantov[86]; reproduced by permission of the American Chemical Society.

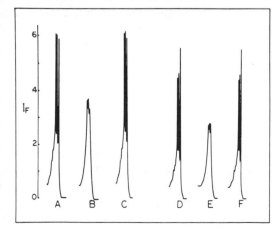

multiplets are not altered. Subsequent annealing steps at temperatures lower than the "new" one used for annealing fail to alter the "new" intensity distribution of the multiplets, as shown in Figure 18.[86] These results indicate that the site modifications that affect the multiplet structure are not destroyed at any temperature markedly lower than that at which the sites in question were originally produced. The easiest way to understand this observation is to postulate that some conformational change in the matrix material (e.g., rotational conformers of alkane chains in n-alkanes[30,31]) is responsible (Sec. C.1). At present, this argument is rather speculative and further study is required to understand the origin of the temperature-dependent multiplet structure and memory effects in vapor-deposited Shpol'skii matrices.

A Shpol'skii matrix obviously offers the advantage of increased spectral resolution over that observable in conventional vapor-deposited matrices (however, see Sec. E). There are two principal disadvantages of Shpol'skii matrices in MI fluorometry. First, the fluorescence blank in organic solvents is larger and more difficult to decrease than that characteristic of conventional matrices. A matrix material such as N_2 or argon, which has a low boiling point, can be effectively purged of most fluorescent contaminants by passing it through a series of cold traps at liquid nitrogen temperature. Nevertheless, detection limits in both conventional and Shpol'skii MI fluorometry are blank limited, with the blank in the former case usually being smaller than in organic matrices. As a result, when continuum-source excitation is used, detection limits for MI fluorometry tend to be lower in conventional matrices than in Shpol'skii matrices.

A second important disadvantage of Shpol'skii matrices (either frozen solution or vapor deposited) is the fact that they are less "inert" than conventional MI media. One consequence of this fact is that photochemical decomposition of analytes tends to proceed much more efficiently in organic matrices. Such an occurrence acts as a form of "gain drift,"[87] which decreases measurement precision. In addition, for polar solutes such as phenols or heterocycles, Shpol'skii matrices often produce spectral resolution inferior to that obtainable in more nearly "inert" conventional matrices; for an example see Figure 19. Presumably this effect arises from matrix-solute interactions, which for Shpol'skii matrices may include weak hydrogen bonding.

Because Shpol'skii spectra can be obtained either by frozen-solution or matrix isolation procedures, one may also inquire into their relative advantages and shortcomings. The principal advantage of frozen-solution methods is that they are simpler (and usually faster) than matrix isolation. The apparatus required for frozen-solution spectroscopy is also less expensive than that employed in MI. The principal advantage of MI is that if the experiment is performed properly, aggregation effects and the other artifacts that can arise in frozen-solution experiments (Sec. C.2) are absent. One consequence of this fact is that the linear dynamic range for MI fluorometry in conventional media is retained or even extended in Shpol'skii matrices. For example, using dye laser excitation (see Sec. D.3), analytical calibration curves linear from subpicogram detection limits to analyte quantities in the microgram range have been obtained.[85] The authors are unaware of any instance in which linear behavior over such a wide concentration range has been reported in frozen-solution matrices (Shpol'skii or otherwise). An additional advantage of MI over frozen solutions is that in MI *any* suitable organic solvent can be used, irrespective of the solubility properties of the sample in that solvent. In contrast, solvents for frozen-solution spectroscopy must obviously dissolve the sample constituents at room temperature.†

3. Laser-Induced Matrix Isolation Fluorometry

An important aspect of Shpol'skii spectroscopy in either frozen solutions or vapor-deposited matrices is that maximum selectivity can be achieved only if narrow-bandwidth exciting light is employed. As noted in Sec. C.1, both excitation and emission spectra undergo narrowing when measured in Shpol'skii matrices. Therefore selective excitation of fluorescence from individual compounds in complex mixtures is feasible if the excitation source exhibits a bandpass of 10 cm^{-1} or less. In principle, either a continuum source–monochromator combination or a tunable laser could be used for excitation; in practice, achievement of useful incident photon fluxes under these conditions is generally easier if a laser is employed. The use of dye laser excitation in Shpol'skii MI fluorometry has recently been reported by Maple *et al.*[85] Of particular interest is the capability for selective excitation offered by laser illumination. For example, Figure 20 shows a laser-excited fluorescence spectrum in *n*-heptane of the same coking plant water chromatographic fraction whose fluorescence spectrum obtained by continuum-source excitation in a nitrogen matrix is shown in Figure 13. The excitation wavelength used to obtain the spectrum in Figure 20 is the optimum wavelength for excitation of Shpol'skii fluorescence of perylene in *n*-heptane. In Figure 21 is shown a fluorescence spectrum, obtained at the same excitation wavelength, for pure perylene matrix isolated in heptane. Comparison of Figures 20 and 21 shows a remarkable resemblance. It should be recalled that the coking plant water sample contains at least 25 fluorescent compounds, yet only perylene fluorescence is observed in

† For example, site-selection fluorescence spectra of hydroxynaphthalenes can be observed by MI in perfluoro-*n*-hexane, a solvent in which these analytes are virtually insoluble at room temperature [J. R. Maple and E. L. Wehry, *Anal. Chem.* **53**, 266 (1981)].

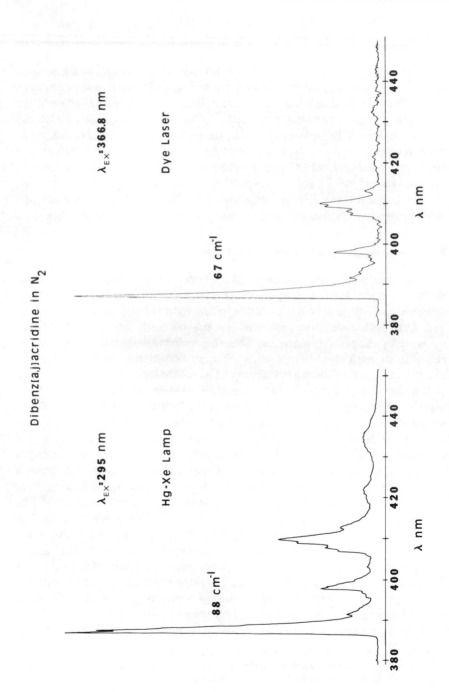

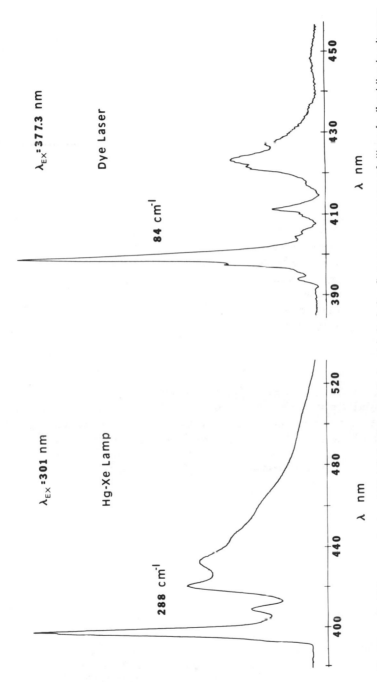

FIGURE 19. (Left) Lamp-excited and (right) dye-laser-excited matrix-isolation fluorescence spectra of dibenz[a,j]acridine in nitrogen (top) and n-heptane (bottom) matrices. Note that N₂ matrices yielded higher spectral resolution than heptane, irrespective of the source, and that in both matrices laser excitation produced higher resolution than lamp excitation.

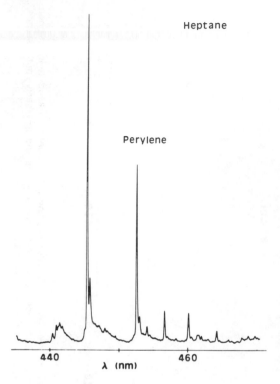

Heptane

Perylene

440   λ (nm)   460

FIGURE 20. Dye-laser-excited fluorescence spectrum of absorption chromatography fraction from coking plant water sample, matrix isolated in *n*-heptane. Excitation wavelength; 409.5 nm; excitation band-wdith; 0.02 nm. Compare with Figure 13. Although this sample contains at least 25 fluoro-phores, the spectrum is virtually identical to that obtained for pure perylene under the same conditions (cf. Figure 21).

heptane if the proper excitation wavelength and excitation bandwidth are employed. Another example of the same phenomenon, involving this same sample, is shown in Figure 22. In this example, laser-excited fluorescence spectra of the sample and pure benzo[*a*]pyrene in heptane are compared

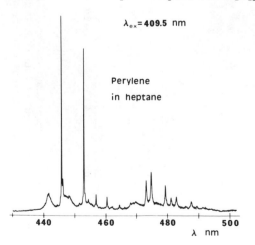

$\lambda_{ex} = 409.5$ nm

Perylene

in heptane

440   460   480   500

λ nm

FIGURE 21. Fluorescence spec-trum of perylene matrix isolated in *n*-heptane, excited by dye laser. Excitation conditions are the same as for Figure 20. From Maple, Wehry, and Mamantov[84]; reproduced by permission of the American Chemical Society.

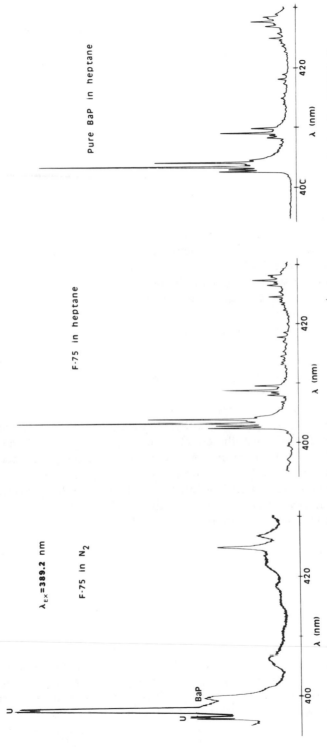

FIGURE 22. Laser-excited matrix-isolation fluorescence spectra ($\lambda_{ex} = 389.2$ nm) of coking plant chromatography fraction (F-75) in N_2 (left) and *n*-heptane (center) matrices. The spectrum at the far right is that of pure benzo[*a*]pyrene in a *n*-heptane matrix. Note the similarity of the center and right spectra, indicating the high selectivity with which BaP fluorescence can be excited in the multicomponent sample. *U* in the N_2 spectrum denotes an unidentified sample constituent. From Maple, Wehry, and Mamantov[85]; reproduced by permission of the American Chemical Society.

with each other and with a spectrum of the coking plant water (laser excited) in a nitrogen matrix. Again, the spectra of the sample and of the pure compound in question (BaP) are virtually identical when a Shpol'skii matrix and laser excitation are used. Analogous results are not obtained in nitrogen matrices, nor are similar results obtainable in any matrix when continuum-source illumination is employed. The possibilities for selective excitation of fluorescence from individual compounds in complex mixtures by laser-excited fluorescence in Shpol'skii matrices are indeed intriguing, and this technique is presently being applied to the characterization of coal liquids and other very complex samples.[88]

In principle, it might seem reasonable to expect improvement of detection limits for fluorescence in Shpol'skii matrices by laser excitation. Because the widths of individual bands in MI fluorescence spectra obtained in Shpol'skii matrices are much less than the bandwidths of the output of typical lamp–monochromator combinations operated under realistic analytical parameters, efficient absorption of the output of a continuum source cannot be achieved in Shpol'skii matrices. Thus, the "fluorescence advantage"[89] is not exploited effectively. In contrast, it is relatively easy to obtain a bandwidth from a dye laser comparable to or less than that of the absorption bands in a Shpol'skii spectrum. Unfortunately, however, most analytes that exhibit useful Shpol'skii fluorescence absorb most strongly in the fairly deep UV (320 nm or lower wavelengths), wherein fundamental dye laser output is not now available and it is necessary to resort to second-harmonic generation or other nonlinear processes to achieve tunable laser output. Hence, the effective optical power per unit wavelength interval achievable by many due lasers in the UV is actually exceeded by that which can be obtained from a xenon–mercury lamp of high power. Hence, the anticipated "sensitivity" advantage of laser over lamp excitation of Shpol'skii matrices often fails to materialize.[85] In a matrix such as nitrogen, it is common for laser excitation to produce detection limits inferior to those obtainable via lamp excitation, often by several orders of magnitude; in this regard, the lower resolution achievable in conventional matrices trades off very nicely for an improvement in sensitivity. For the near future, therefore, the principal use of laser excitation in MI fluorometry will be to achieve enhanced analytical selectivity, and these prospects have barely begun to be elucidated.

Another application of laser technology to analytical fluorometry is time-resolved spectroscopy (see Chapters 1 and 2). In some cases, time-resolved fluorometry can provide additional improvement in the selectivity of MI fluorometric analysis. The possibilities here are most interesting when the analyte is encountered in the presence of a spectral interferent present in large excess, wherein the interferent exhibits a shorter fluorescence decay time than the analyte. A realistic example of such a situation is the pair of isomeric polycyclic aromatic hydrocarbons benzo[a]pyrene and benzo[k]fluoranthene. The former is a notorious carcinogen; the latter is much less hazardous but often appears in energy-related samples in large excess over benzo[a]pyrene. In conventional matrices, the fluorescence spectra of these two compounds overlap seriously; as shown in Figure 23, it is virtually impossible to detect (much less quantify) benzo[a]pyrene in the presence of a substantial excess of benzo[k]fluoranthene. It happens, however, that their fluorescence decay times are significantly different (BaP, 13 nsec; BkF, 78 nsec); hence, time-resolved detection of BaP is relatively straightforward (as also shown in Figure 23).[90] Quantitation of BaP in the presence of excess BkF is thus greatly facilitated in conventional matrices by time-resolved fluorometry. However, it must also be pointed out that an alternative procedure for discriminating between these two compounds is to do so in the spectral, rather than the time, domain by carrying out the MI fluorometric measurement in a Shpol'skii matrix.[85] For most compounds that cannot be resolved spectrally, it turns out that their decay times are so similar that meaningful resolution in the time domain is also unfeasible. Another advantage of time resolution, however, is more general. As noted

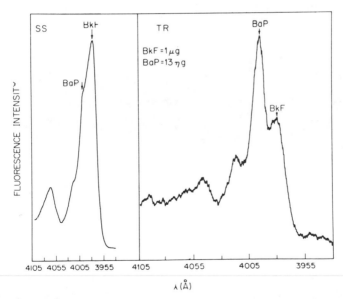

FIGURE 23. Steady-state (left) and time-resolved (right) matrix-isolation fluorescence spectra of two-component mixture (benzo[k]fluoranthene, 1 μg; benzo[a]pyrene, 13 ng). The time delay in the time-resolved spectrum was 90 nsec. From Dickinson and Wehry[90]; reproduced by permission of the American Chemical Society.

previously, background luminescence is observed from all frozen-solution and vapor-deposited matrices; moreover, these materials often fail rather miserably one of the criteria of an "ideal solvent," in that they tend to exhibit strong Rayleigh scattering of the incident light. Complex real samples such as coal liquids also tend to exhibit a low-level broadband background luminescence upon which the structured emission bands of the analyte species are superimposed. This "spectral garbage" can be nicely reduced by time resolution, as shown in Figure 24.

A particularly useful characteristic of cryogenic media (especially vapor-deposited matrices) is that fluorescence decay times are often much longer than those observed in fluid media. Hence, particularly if it is not desired to measure accurate fluorescence decay curves, rather simple apparatus can be employed in time-resolved cryogenic fluorometry. Thus, whereas the time-resolved spectra shown in Figure 23 were obtained using a complicated spectrometer based on a mode-locked cavity-dumped argon-ion laser,[90] those shown in Figure 24 were measured with a simpler spectrometer using a nitrogen-pumped dye laser as a source. Although the latter laser produces "long" pulses (5–10 nsec), they are satisfactory for most "background-reduction" applications of time-resolved fluorometry in low-temperature media because fluorescence decay times smaller than 10 nsec are rather rarely observed in such media.

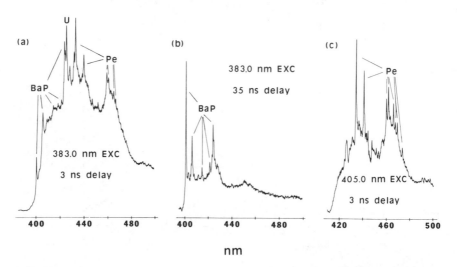

FIGURE 24. Pulsed dye-laser-excited matrix-isolation fluorescence spectra of coal-derived material (SRC-1) in argon matrices at excitation wavelengths and delay times, respectively, of (a) 383.0 nm and 3 nsec; (b) 383.0 nm and 35 nsec; (c) 405.0 nm and 3 nsec.

4. Combination of Matrix Isolation with Separation Techniques

As noted above, an advantage of matrix isolation over frozen-solution spectroscopy is that MI can be combined directly with chemical separation procedures. It is important to realize that, irrespective of the spectral resolution attainable by a technique, there is reached a point at which the complexity of real samples defies analysis. For example, coal liquids may contain hundreds of compounds, and the number of these that may fluoresce may be quite large. Hence spectroscopic resolution of such a sample may not be possible in a reasonable time unless some sort of sample fractionation is first performed. It is desirable, when feasible, to couple the separation step to the spectroscopic measurement, to create what Hirschfeld[91] terms a "hyphenated" method. Because the sample preparation procedure in MI requires that the original sample be vaporized, coupling MI with separation methods proceeding by the vapor phase is possible. One very simple example involves exploiting the fact that the various compounds present in a complex sample have different boiling points. Hence, if the Knudsen cell in a MI experiment is operated in such a way that the temperature is either periodically "stepped" to higher values or linearly programmed, then the sample will undergo fractional sublimation or fractional distillation. Coupling this procedure with the use of a cryostat equipped with several different surfaces allows the MI sample preparation step itself to serve as a fractionation method.[92] It is not realistic to expect such a simple procedure to produce separations of high quality in complex samples, but the technique can serve to "decompose" samples of moderate complexity into individual multicomponent samples that can be characterized by selectively excited MI fluorometry.

An obvious extension of this idea is to couple MI fluorometry directly to gas chromatography. It may initially be argued that such a procedure is an "overkill" in the sense that if the chromatographic separation proceeds with high efficiency, high-resolution fluorescence spectra are not really needed. For many complex samples, however, "complete" separations of potentially interfering sample constituents are not achieved and the "fingerprint" information produced by high-resolution spectroscopic detection is then particularly useful. This is, of course, one of the rationales for the combination of gas chromatography with mass spectrometry.

The use of matrix-isolation Fourier-transform infrared spectrometry for gas chromatographic detection has been reported by Reedy and co-workers[93,94] and is also under investigation in this laboratory.[92] Matrix isolation fluorescence spectroscopy is also being explored regarding its utility in gas chromatographic detection; for more detail on this subject, the reader is referred to Chapter 2 in Volume 3 of this series.

E. SITE-SELECTION FLUORESCENCE SPECTROMETRY

1. Fundamentals

We have noted in preceding sections that one way to achieve high spectral resolution in low-temperature fluorometry is to minimize the degree of microenvironmental heterogeneity experienced by the solute in the matrix; this is the basis of the Shpol'skii effect. A principal shortcoming of this approach is the fact that different matrices are optimal for different analytes. An alternative approach to the problem is to use the same matrix for all samples, but to contrive to obtain spectra in which the bandwidths are minimized by selective excitation only of those solute molecules that inhabit very similar matrix sites. Such techniques are described in the literature under various names, including "site-selection spectroscopy," "fluorescence line narrowing," and (probably most accurately) "energy-selection spectroscopy." Kohler[7] has recently reviewed this procedure, and only those aspects most germane to eventual analytical application of the phenomenon are discussed here.

The first reported site-selection experiment in a solid sample was performed by Szabo,[95,96] who described the process of narrowing the 6934-Å emission line of Cr^{3+} in ruby by site-selective excitation at 4.2°K. Spectral linewidths as small as 0.002 cm^{-1} were obtained in those experiments. The principle was first applied to organic molecules dissolved in frozen organic solvents in 1972 by Personov, Al'shits, and Bykovskaya.[97] Subsequent studies by Personov and co-workers[98–101] and others[26,102–117] have established many of the systematics of site-selection spectra and the conditions required for their observation.

The most obvious requirement in a site-selection (SS) experiment is the use of a source whose output exhibits a bandwidth appreciably smaller than that of the inhomogeneously broadened electronic transition of the analyte molecule. An example of the fluorescence band narrowing that can be achieved in favorable circumstances via narrow-bandwidth excitation in non-Shpol'skii cryogenic matrices is shown in Figure 25.[107] A number of other requirements must also be satisfied in order for SS to be observed; these have been summarized by (among others) Kohler[7] and McColgin et al.[105] The most important requirements and (where applicable) the reasons for them are listed below.

(a) If maximum spectral resolution is desired, the excitation bandwidth should not substantially exceed the width of the zero-phonon absorption line of an assembly of solute molecules occupying a well-defined type of site. Therefore, the source bandwidth and the emission monochromator spectra bandpass should be on the order of 0.1 cm^{-1} for optimum results.[7] Studies

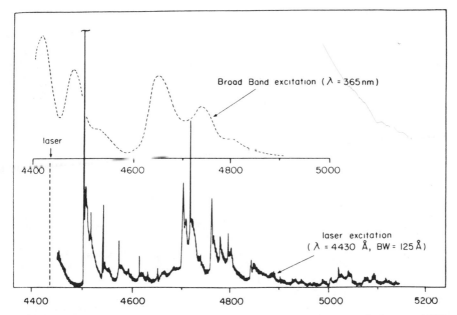

FIGURE 25. Fluorescence spectra of perylene in ethanolic frozen solutions at 4.2°K. (Top) Hg lamp excitation source; (bottom) dye laser excitation source ($\lambda_{ex} = 443.0$ nm; bandpass = 0.0125 nm). From Abram *et al.*[107]; reproduced by permission of the American Institute of Physics.

of SS using excitation bandwidths as small as 0.01 cm^{-1} have been reported.[102] As noted below, the source should be continuously tunable. Either a dye laser, usually fitted with at least one intracavity etalon,[105,107,112,114] or a powerful continuum source dispersed by a high-resolution monochromator[26,108] can be used for excitation. In favorable cases zero-phonon lines having widths less than 1 cm^{-1} are observed in SS fluorescence spectra.[118,119]

Under conditions of very narrow source bandwidth and excitation monochromator bandpass (i.e., low throughout), acquisition times to achieve adequate signal-to-noise ratios in SS fluorometry can become prohibitive.[120] If maximum spectral resolution is unnecessary or unfeasible, fluorescence spectra exhibiting manifestations of site selection can be obtained by the use of larger source bandwidths and/or lower monochromator resolution (e.g., 12 cm^{-1} [109]). In such situations the site selectivity of the experiment is decreased and the observed spectral resolution will be limited by the source bandwidth or monochromator bandpass (whichever is poorer).[110]

(b) The solute in question must usually be excited in its 0–0 absorption region.[101] In addition, it is desirable for the excitation energy to correspond to the zero-phonon line (rather than the phonon wing) of the 0–0 band.[101,107] An example of the appearance of fluorescence spectra of the polycyclic aromatic hydrocarbon pyrene in a 1 : 1 glycerol–ethanol glass as a function of excitation wavelength is ashown in Figure 26.[121] In general, production of vibrationally excited solute molecules should be avoided. In

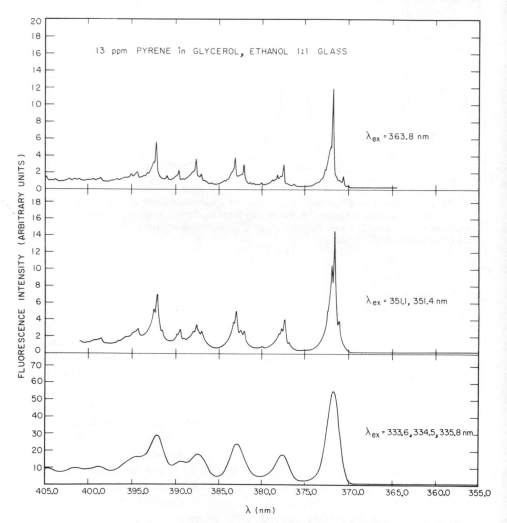

FIGURE 26. Dye-laser-excited fluorescence spectrum of pyrene in glycerol–ethanol frozen-solution matrix at 4.2°K as a function of excitation wavelength. From Brown, Edelson, and Small[121]; reproduced by permission of the American Chemical Society.

that regard, Kohler[7] and Labhart *et al.*[120] have stressed that the term "site selection" is something of a misnomer. Whenever one chooses a narrow excitation bandwidth within the 0–0 absorption region for a compound, one is actually performing an "energy-selection" experiment, wherein a distribution of molecules having very nearly the same purely electronic transition energy is selected. There is no guarantee that the selected molecules all exhibit identical, or nearly identical, lattice sites. Thus, one does not expect all selected molecules to exhibit identical ground- or excited-state vibrational frequencies. Hence, as soon as population of vibrationally excited levels of the electronically excited state occurs, the probability that different vibrational sublevels corresponding to different solute sites will be populated becomes appreciable.[102,107] As the excitation energy is increased further into more congested regions of the absorption spectrum, virtually all band narrowing is lost. Moreover, when electronic excitation produces vibrationally excited solute molecules, the subsequent rapid vibrational relaxation and transfer of thermal energy to the matrix may cause appreciable "local site melting".[7] This will have the effect of altering the site structure of the matrix and, in extreme cases, the initial site selection will be virtually negated.[109,110]

A consequence of the "energy selectivity" of this type of experiment is that dramatic alterations in the appearance of a fluorescence spectrum are brought about by small changes in the excitation wavelength. Not only are the relative intensities of the various emission spectra features very sensitive to the excitation wavelength, but shifts in the band positions also occur. An example of this effect is shown in Figure 27. Excitation into the 0–0 band leaves no opportunity for excited-state vibrational relaxation of the solute, and rigid cryogenic matrices exhibit essentially no reorientation following electronic excitation of the solute (i.e., there really is no such thing as a "Franck–Condon excited state"[122] of the solute in these experiments). Because a narrow energy distribution of solute molecules is excited, it follows that the 0–0 zero-phonon fluorescence line in a SS spectrum should occur at the same wavelength as the exciting radiation.[105] In many practical cases, the 0–0 fluorescence line is partially attenuated by self-absorption or obliterated by Rayleigh scattering of the incident light. Fortunately, various fluorescence transitions to vibrationally excited ground-state molecules often exhibit line narrowing, and these also tend to shift with changes in the exciting wavelength (Figure 27). Observations such as those shown in Figure 27 imply that in the different lattice sites the electronic transition energies of the various solute molecules vary to a much greater extent than the vibrational frequencies.[105] For solutes, or in matrices, wherein this is not the case, molecules having the same 0–0 electronic transition energy may actually occupy a wide range of different sites. In that case, the term "site

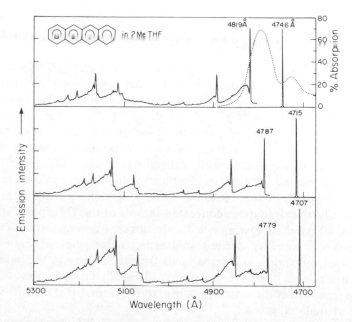

FIGURE 27. Site-selection fluorescence specta of tetracene in 2-methyltetrahydrofuran frozen solutions at 4.2°K as a function of excitation wavelength. The laser line (exciting wavelength) is shown in each spectrum at the extreme right. Note that the spacings between the excitation wavelength and the first major band in the tetracene fluorescence spectrum are virtually equal in the three spectra. The dotted curve in the top spectrum is the absorption spectrum of tetracene in the same glass at the same temperature. From McColgin, Marchetti, and Eberly[105]; reproduced by permission of the American Chemical Society.

selection" is especially inappropriate and the extent of "vibrational inhomogeneous broadening" will be large and effective line narrowing will be difficult to achieve.

Among other things, all of this suggests that the source in SS fluorometry not only must exhibit a narrow bandwidth, but both the bandwidth and the wavelength itself should be highly reproducible for analytical studies, particularly for "fingerprinting."

(c) Site-selection fluorescence spectra are usually observed only at very low temperatures (5°K or less).[7,98,110,121] As noted in Sec. C.1, the intensity ratio of the zero-phonon line to the phonon sideband (Debye–Waller factor) is very dependent on the temperature of the sample.† Evidence for the existence of different sites in frozen-solution matrices at temperatures as

† However, instances are occasionally observed in glassy frozen solutions in which the zero-phonon intensity actually increases with increasing temperature over a limited range of temperatures.[113]

high as 240°K has been reported,[108] but much lower temperatures must be employed to observe an appreciable fraction of the fluorescence of a solute molecule in the zero-phonon line instead of the phonon wing. Moreover, the distribution of sites within the matrix must remain constant on the fluorescence time scale. In the absence of a site distribution function that remains virtually constant, the concept of site selection becomes virtualy meaningless. Hayes and Small[113,114] have suggested two mechanisms for site interconversion of electronically excited solute molecules in solid matrices; both of these processes (thermal activation and phonon-assisted tunneling) are predicted to exhibit a strong temperature dependence.

(d) To observe SS spectra, it is essential that solute molecules occupying different sites do not "communicate" with each other. An obvious mechanism for such communication is intermolecular electronic energy transfer. If appreciable transfer of electronic energy from molecules occupying a narrow distribution of sites occurs to other solute molecules that had not been "selected" for direct excitation, the desired fluorescence band narrowing will not be observed. That SS fluorescence spectra have been observed for a rather wide variety of organic solutes in a variety of frozen-solution and vapor-deposited matrices indicates that energy transfer between sites is often negligible. Clearly, it is crucial for solute aggregation to be negligible; the average distance between sites should be at least on the order of the critical transfer distance for Förster long-range transfer. Transfer of electronic energy between sites in solid samples can be detected by time-resolved measurement of the bandwidth of the emitted fluorescence.[123]

(e) At a given temperature, the extent of band narrowing for a given solute is different in different matrices. For example, at 2°K the polycyclic aromatic hydrocarbon tetracene exhibits excellent band narrowing in a diethyl ether–isopropanol frozen solution but under the same conditions much less dramatic band narrowing in ethanol–methanol matrices.[114] Such observations are related to the extent to which the electronic transition of the solute is coupled to phonons of the lattice (see Sec. C.1), which cannot be necessarily predicted *a priori* (especially for large solute molecules and mixed solvent systems). Thus a certain amount of trial and error may be required to identify the "optimum" matrix for SS spectroscopy of a particular compound. One is therefore not totally free of the solvent-dependence problem (although it is usually much less severe in SS than in Shpol'skii fluorometry).

(f) The Stokes shift (i.e., the difference in wavelength or wave number between the 0–0 absorption and fluorescence bands) observed under conditions of broadband excitation has been shown theoretically to vary inversely with the Debye–Waller factor.[105] This is another way of saying that the

Stokes shift observed under broadband excitation arises from the phonon sidebands in the absorption and fluorescence spectra, averaged over the distribution of sites in the sample. Thus, a rough prediction of the relative intensity of the zero-phonon line to that of the phonon wing under narrow-bandwidth excitation can be made by measuring the Stokes shift, in the same solvent at the same temperature, with broadband excitation. For example, tetracene (for which SS fluorescence spectra in 2-methyltetrahydrofuran at 4.2°K are shown in Figure 27) exhibits a Stokes shift of $\sim 87 \text{ cm}^{-1}$ in that solvent at 4.2°K under broadband excitation. For a molecule exhibiting a larger Stokes shift, the relative intensity of the zero-phonon line to the phonon sideband will tend to be smaller; e.g., the dye resorufin in ethanol–methanol glass at 4.2°K exhibits a Stokes shift of $\sim 360 \text{ cm}^{-1}$ and the SS fluorescence spectrum shown in Figure 28.[105] For compounds exhibiting even larger Stokes shifts in a certain solvent at a given temperature, attempts to obtain appreciable line narrowing under narrow-bandwidth excitation may become a losing proposition.

(g) While SS fluorescence has been observed for a sizable number of unsubstituted polycyclic aromatic hydrocarbons, those aromatics possessing substituents (such as alkyl, phenyl, or halogens) appear less susceptible to line narrowing by narrow-bandwidth excitation.[102,108] Relatively few attempts to observe SS fluorescence spectra of highly polar solutes have been reported; however, it is encouraging that SS has been observed for various porphyrins[100,112] and for at least one ionic organic compound (the anionic dye resorufin[104]).

These observations may derive in part from the fact that solute molecules for which electronic excitation produces large alterations in molecular geometry are relatively unlikely to exhibit extensive line narrrowing.[7]

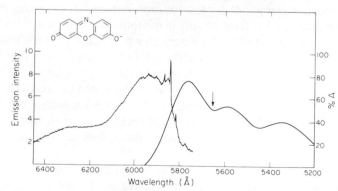

FIGURE 28. Site-selection fluorescence spectrum of resorufin in ethanol–methanol frozen solution at 4.2°K. From McColgin, Marchetti, and Eberly[105]; reproduced by permission of the American Chemical Society.

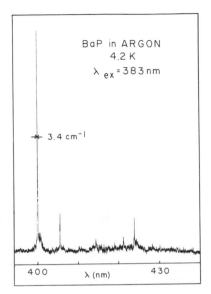

FIGURE 29. Site-selection matrix isolation fluorescence spectrum of benzo[a]pyrene in vapor-deposited argon matrix at 4.2°K. The bandwidth of the major feature (3.4 cm⁻¹) is limited by the resolution of the 1-m grating monochromator used to obtain the spectrum.[124]

Hence, very rigid molecules (such as unsubstituted polycyclic aromatic hydrocarbons) constitute the most favorable species for SS spectroscopy; substituents are likely to decrease the effective "rigidity" of an aromatic solute molecule.

(h) Fluorescence excitation spectra containing very sharp lines can be measured by the narrow-bandwidth laser approach, but only within ~1500 cm⁻¹ of the 0–0 absorption band.[110,119] This fact has been attributed to local site melting produced by energy released to the matix by rapid vibrational relaxation in the solute.

(i) Finally, although all examples of SS described above have been observed in frozen-solution media, site selection can also be observed by matrix isolation in vapor-deposited rare gases (Sec. D). An example of a SS fluorescence spectrum of benzo[a]pyrene in a vapor-deposited argon matrix is shown in Figure 29.[124]

One may ask whether site-selective excitation produces line narrowing in phosphorescence, as well as fluorescence, spectra. The answer, in general, is negative when narrow-band excitation of the first excited singlet state (S_1^*) is employed under conditions in which line-narrowed fluorescence spectra are observed.[77,110,125–129] An example comparing the fluorescence and phosphorescence spectra of 1-chloronaphthalene in ethanol matrices at the same excitation wavelength (Figure 30) dramatically illustrates the lack of observed band narrowing in phosphorescence.[110]

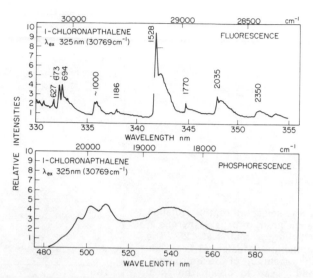

FIGURE 30. Fluorescence (top) and phosphorescence (bottom) spectra of 1-chloronapthalene in ethanolic frozen solutions at 10°K. Note that both spectra were excited at the same wavelength (325 nm); the fluorescence spectrum exhibits band narrowing but the phosphorescence does not. From Cunningham *et al.*[110]; reproduced by permission of North-Holland.

Several possible explanations for this and similar observations have been proposed. One plausible argument invokes loss of site selectivity in phosphorescence due to energy transfer (which, as noted previously, is not a general problem in SS fluorescence but could be important in phosphorescence because of the long mean lifetimes of triplet states, especially in cryogenic solids). However, it has been observed on a number of occasions[125,126,128] that line-narrowed phosphorescence spectra can be obtained by narrow-bandwidth excitation in the singlet–triplet (i.e., $S_0 \rightarrow T_1^*$) absorption region. The conditions for band-narrowing in fluorescence and phosphorescence generally appear to be the same, with the ultimate emitting state being the one that must initially be produced by narrow-bandwidth excitation. Such observations virtually eliminate the possibility that energy transfer between molecules of the same compound occupying different sites can be a general process for loss of site selectivity in phosphorescence when S_1^* is initially excited. Instead, it appears that there fails to exist anything resembling a strong correlation between $S_0 \rightarrow S_1^*$ and $S_0 \rightarrow T_1^*$ transition energies in frozen solutions at 4°K. (No attempts to observe SS phosphorescence in vapor-deposited matrices have yet been reported.) That is, a given ensemble of solute molecules having nearly identical $S_1^* \rightarrow S_0$ transition energies are likely to exhibit disparate $T_1^* \rightarrow S_0$

energies and vice versa.[125-128] Once again, it must be recalled that site selection is really not an accurate description of these experiments.

It should also be recalled that in most molecules the ground vibrational level of T_1^* is substantially lower in energy than the lowest vibrational level of S_1^*. Hence, when $S_1^* \to T_1^*$ intersystem crossing occurs, a significant amount of energy is released to the matrix prior to emission . Significant local site melting may therefore occur,[110,126] thus further reducing the probability that SS phosphorescence will be observed when excitation is into the singlet manifold. On this basis one would expect solute molecules having very small $S_1^*-T_1^*$ energy separations to exhibit some manifestations of phosphorescence line narrowing when excited into the lowest vibrational level of S_1^*.

From the analytical viewpoint, the fact that phosphorescence band narrowing is generally observable only when singlet–triplet excitation is employed renders SS phosphorescence a losing proposition. Even in heavy-atom solvents such as 1-bromobutane, solute concentrations on the order of 10^{-2}–10^{-3} M are required to obtain adequate signal-to-noise ratios in SS phosphorometry.[126] It is difficult to envision any analytical future for the technique unless a more efficient measns of excitation is discovered.

An additional study that has intriguing spectroscopic (though not necessarily analytical) implications indicates that SS phosphorescence in one solute can be sensitized by narrowband $S_0 \to T_1^*$ excitation of another solute.[129] For example, narrow-band excitation of 1-iodonaphthalene in an EPA glass at 4.2°K containing perfluoronaphthalene produces narrow-band phosphorescence of the latter. Hence, in this case, the site selection is preserved in the energy-transfer process. Further experimental studies along these lines may assist in understanding the extent to which narrow-bandwidth excitation really effects site-selective excitation in different types of matrices.

2. Analytical Applications

At the time of this writing (April 1980), the only published analytical study of SS fluorometry is that of Brown et al.,[121] who have examined the fluorescence spectra of several polycyclic aromatic hydrocarbons in glycerol–water and glycerol–water–dimethyl sulfoxide[133] glassy frozen-solution matrices at 4.2°K. Analytical calibration curves for quantitation of pure anthracene and pyrene were linear over three decades in concentration, with detection limits of 1 ppb or less and precision on the order of 8% relative. The procedure has been applied to determination of several polycyclic aromatic hydrocarbons in coal liquefaction products, with promising results.[133] It has been claimed that quantitative SS fluorometry in glassy

frozen solutions can be carried out without use of either internal standardization or standard addition.[121] Whether this proves to be a real or an
illusory advantage of the procedure remains to be seen; many complex real
samples must be "spiked" with some sort of standard prior to analysis in
order to estimate the efficiency of recovery or cleanup steps in actual
practice.[130]

The ability to obtain very-high-resolution molecular fluorescence
spectra of many different molecules in a single sampling medium, using
either frozen-solution[121] or matrix isolation[124] techniques, is an
extremely attractive feature of SS fluorometry, and applications of the
procedure to complex real samples should be very fruitful. However, the
technique is not without real and potential disadvantages, several of the
most obvious being itemized below.

(a) The requirement that spectra be measured at liquid helium
temperatures involves both cost and inconvenience, as compared with
closed-cycle refrigerators[79] or liquid nitrogen cryostats.

(b) Use of a tunable laser for excitation is absolutely necessary to realize
the full analytical potential of SS fluorometry. As noted previously, this is
also true for both Shpol'skii and matrix isolation fluorometric techniques, so
this does not constitute a unique disadvantage of site selection.

(c) The quantitative precision of a site-selection experiment depends
upon the extent to which the "site distribution" for a given analyte is
reproduced from one sample to the next. Only a fraction of the molecules of
a given solute is chosen for possible excitation in a band-narrowing experiment; if that fraction varies markedly between samples, the quantitative
reproducibility may be inferior to that of other low-temperature
fluorometric procedures. It is very encouraging in this regard that preliminary studies[121] in frozen solutions do not indicate the precision of
quantitative analyses of pure compounds by SS fluorometry to be appreciably worse than that of more conventional low-temperature fluorometric
procedures. This point obviously requires and certainly will receive
continued attention.

(d) A phenomenon termed "nonphotochemical hole
burning"[113,114,131,132] could conceivably cause problems in quantitative SS
fluorometry. When narrow-bandwidth laser excitation in a matrix at very
low temperatures (10°K or less) is carried out, the absorption spectrum of
the absorbing solute may, after a period of continued exposure to the laser
beam, begin to exhibit narrow gaps, the depths of which increase with
continued illumination. This phenomenon can be observed for solutes that
exhibit no detectable photodecomposition under the experimental conditions (hence the name "nonphotochemical" hole burning). The resulting

alterations in the absorbance of the analyte could produce drift in the measured fluorescence intensity. The phenomenon appears to be related to site interconversions of solute molecules while in electronically excited states.[113,114] It appears that the analytical significance of hole burning will generally be negligible in both frozen-solution[121] and vapor-deposited matrices in SS experiments if the time interval during which the sample is exposed to the laser beam at a particular wavelength is reasonably small.

(e) Finally, the sensitivity of SS is likely to be somewhat inferior to that of conventional frozen-solution or matrix-isolation fluorometric techniques for two reasons. First, the SS procedure, by definition, senses only a fraction of the analyte molecules present in a sample, whereas broadband excitation includes virtually 100% of the molecules of the analyte. Second, achievement of the desired spectral resolution in SS requires use of high-resolution monochromators and narrow slit widths (hence, low throughput), and the dye laser used for excitation may also have to be equipped with one or several intracavity etalons to reduce the bandwidth of its output (which exacts a considerable cost in power output). Therefore SS seems best suited to analytical situations in which selectivity rather than sensitivity is the prime requirement. For many real samples selectivity is indeed the major analytical need, and it is therefore likely that SS fluorometry will satisfy a genuine analytical need.

In this context, Labhart *et al.*[120] have noted that the observed fluorescence spectrum of a molecule can be regarded as a convolution of the pure electronic transition with the pattern of vibrational transitions. If one assumes that the vibrational spacings are essentially the same for molecules occupying different sites in a cryogenic solid, a deconvolution technique using rather conventional Fourier-transform computational procedures can be employed to virtually eliminate the multiple-site structure from a fluorescence spectrum measured with broadband excitation. This recently reported procedure, which can proceed under spectral resolution conditions much less stringent than those required in "conventional" site-selection fluorometry, has very interesting implications for the performance of practical analytical SS studies.

F. SUPERSONIC JET FLUORESCENCE SPECTROSCOPY

All techniques considered thus far in this chapter require that the analytes be embedded in some sort of cryogenic solid matrix. Very-high-resolution molecular electronic spectra can be obtained by an alternative low-temperature approach, wherein supersonic expansion of a gas through an orifice into a vacuum is used to cool the gas to very low translational temperatures. As the gas is expanded, collisions occur between the gas molecules; the result is a dramatic decrease in the range of translational velocities exhibited by the molecules of the gas. The distribution of molecular velocities can, by this technique, be reduced to the extent that translational temperatures of less than $1°K$ (and on occasions less than $0.1°K$) can be achieved. Fluorescence excitation and emission spectra obtained by this "supersonic jet" procedure have been studied in detail by (among others) Levy and Jortner and their respective groups; Levy and co-workers have recently reviewed the principles and experimental practice of supersonic jet spectroscopy.[134]

After expansion of the gas has been completed, its density is very low and the frequency of molecular collisions is correspondingly small. To the extent that collisions do take place, formation of dimers (or "complexes," if the colliding molecules are different) causes the release of energy to the translational degrees of freedom of the gas, limiting the extent of translational cooling that can be achieved. It is therefore highly desirable that the expanded gas consist largely of atoms or molecules that are chemically very inert; helium is the most popular gas for this purpose because it is believed that the He_2 dimer does not have a bound state. If it is desired to examine the electronic spectrum of some other molecule, that species is normally diluted with helium (or another "inert" gas). In effect, helium serves as the "matrix," and the experiment may be visualized (crudely!) as a form of gas-phase matrix isolation. Helium-to-solute mole ratios on the order of 10^4–10^6 are rather typical in supersonic jet spectroscopic studies of large molecules.†

An example of the spectral resolution that can be achieved in this manner is shown in Figure 31, which shows the fluorescence excitation spectrum of a large polycyclic aromatic hydrocarbon (pentacene) seeded into argon and expanded through a 150-μm nozzle. Effective exploitation of the spectral resolution that can be obtained in these experiments requires laser excitation of fluorescence.[134]

No analytical applications of the supersonic jet principle have yet been reported. It has been proposed[137] that design of a high-resolution spectroscopic gas chromatography detector based on the supersonic jet principle should be feasible; however, the engineering problems asociated with the design and demonstration of a practical GC detector based on this approach are rather formidable and as of this writing (April 1980) do not appear to have been fully overcome. (This approach to chromatographic detection by

---

† For polar molecules (e.g., s-tetrazine[135]), van der Waals "complexes" with helium can be detected in their supersonic jet spectra.

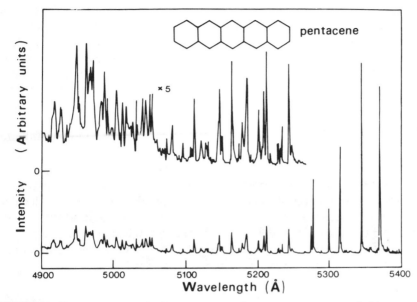

FIGURE 31. Fluorescence excitation spectrum of pentacene vapor seeded into argon and then cooled by supersonic expansion through a nozzle. From Amirav, Even, and Jortner[136]; reproduced by permission of North-Holland.

fluorescence spectroscopy is discussed in greater detail in Chapter 2 of Volume 3.) Despite the experimental difficulties, it seems likely that analytical utilization of the very high resolution obtainable in supersonic jets will eventually be achieved. A realistic assessment of the advantages and shortcomings of the procedure (as compared with those of the solid-matrix techniques discussed in preceding sections of this chapter) cannot yet be undertaken.

G. CONCLUSION

In this chapter, we have examined several procedures, all requiring use of cryogenic temperatures, for the acquisition of molecular fluorescence spectra exhibiting very high resolution. These techniques range from well-established procedures (the Shpol'skii effect) to procedures that have as of yet received no demonstration of analytical practicality (the molecular jet approach). The ultimate objective of research in this broad area is the development of fluorometric procedures that can maximize the "fingerprint" information inherently present in molecular luminescence and excitation spectra and that can provide accurate quantitative analytical data for individual components of complex mixtures with minimal or no sample pretreatment. Indeed, the fluorometric analysis of multicomponent samples represents one of the principal areas of research endeavor in analytical molecular luminescence spectrometry at the present time, and a number of different approaches (which may be competitive or complementary, depending upon the nature of the specific sample under investigation) to this general problem are currently under active investigation; many of these approaches are discussed in other chapters of this book. The ultimate role of the various low-temperature fluorometric techniques in complex-mixture characterization remains to be established, but many of the results obtained to date are extremely encouraging. Moreover, the "cryogenic" requirements of low-temperature spectroscopy can no longer be regarded as a significant experimental obstacle, and lasers suitable for use in molecular fluorescence spectrometry are now commercially available at costs that, while still high, are likely to decrease in the future. It is therefore anticipated that analytical research and analytical utilization of low-temperature fluorescence spectrometry will continue to develop.

ACKNOWLEDGMENTS

Research dealing with matrix isolation fluorescence spectrometry performed in this laboratory has been supported by Contracts RP-332 and RP-1307 with the Electric Power Research Institute. Our studies of laser-induced matrix isolation fluorescence spectrometry have been supported by Grants MPS75-05364 and CHE77-12542 from the National Science Foundation. We would also like to acknowledge the indispensable contributions of the graduate students and postdoctoral associates who have performed research dealing with various aspects of matrix isolation fluorescence spectrometry in this laboratory: Verne Biddle, Paul Bilotta, Bill Carter, Vince Conrad, Dick Dickinson, Randy Gore, Jon Maple, Sumaria Mohan, Mille Perry, Robert Stroupe, and Pari Tokousbalides.

REFERENCES

1. C. A. Parker, *Photoluminescence of Solutions* (American Elsevier, New York, 1968), pp. 89–92.
2. T. Vo-Dinh and J. D. Winefordner, *Appl. Spectrosc. Rev.* **13**, 261 (1977).
3. J. D. Winefordner, W. J. McCarthy, and P. A. St. John, *Methods Biochem. Anal.* **15**, 367 (1967).
4. Reference 1, pp. 220–234.

5. A. A. Lamola, in *Energy Transfer and Organic Photochemistry*, A. A. Lamola and N. J. Turro, eds. (Interscience, New York, 1969), pp. 17–43 and 50–52.
6. A. H. Alwattar, M. D. Lumb, and J. B. Birks, in *Organic Molecular Photophysics*, Vol. 1, J. B. Birks, ed. (Wiley, London, 1973), p. 403.
7. B. E. Kohler, in *Chemical and Biochemical Applications of Lasers*, Vol. 4, C. B. Moore, ed. (Academic Press, New York, 1979), p. 31.
8. B. Meyer, *Low Temperature Spectroscopy* (American Elsevier, New York, 1971), p. 52.
9. J. P. Lemaistre and A. H. Zewail, *Chem. Phys. Lett.* **68**, 296 (1979).
10. E. V. Shpol'skii, A. A. Il'ina, and L. A. Klimova, *Dokl. Akad. Nauk SSSR* **87**, 935 (1952).
11. E. V. Shpol'skii, *Sov. Phys. Usp.* **6**, 411 (1963).
12. A. Colmsjö and U. Stenberg, *Chem. Scr. (Sweden)* **9**, 227 (1976).
13. T. N. Bolotnikova, *Opt. Spectrosc.* **7**, 138 (1959).
14. J. J. Dekkers, G. P. Hoornweg, G. Visser, C. Maclean, and N. H. Velthorst, *Chem. Phys. Lett.* **47**, 357 (1977).
15. G. F. Kirkbright and C. G. de Lima, *Chem. Phys. Lett.* **37**, 165 (1976).
16. A. Colmsjö and U. Stenberg, *Chem. Scr. (Sweden)* **11**, 220 (1977).
17. M. Lamotte and J. Joussot-Dubien, *J. Chem. Phys.* **61**, 1892 (1974).
18. C. Pfister, *Chem. Phys.* **2**, 171 (1973).
19. R. N. Nurmukhametov, *Russ. Chem. Rev.* **38**, 180 (1969).
20. K. Palewska and Z. Ruziewicz, *Chem. Phys. Lett.* **64**, 378 (1979).
21. A. M. Merle, M. Lamotte, S. Risemberg, C. Hauw, J. Gaultier, and J. P. Grivet, *Chem. Phys.* **22**, 207 (1977).
22. M. Lamotte, A. M. Merle, and S. Risemberg, *J. Lumin.* **18–19**, 505 (1979).
23. A. M. Merle, W. M. Pitts, and M. A. El-Sayed, *Chem. Phys. Lett.* **54**, 211 (1978).
24. A. M. Merle, M. F. Nicol, and M. A. El-Sayed, *Chem. Phys. Lett.* **59**, 386 (1978).
25. T. Vo-Dinh and U. P. Wild, *J. Lumin.* **6**, 296 (1973).
26. T. Vo-Dinh, U. T. Kreibich, and U. P. Wild, *Chem. Phys. Lett.* **24**, 352 (1974).
27. P. M. Saari and T. B. Tamm, *Opt. Spectrosc.* **40**, 395 (1976).
28. W. M. Pitts, A. M. Merle, and M. A. El-Sayed, *Chem. Phys.* **36**, 437 (1979).
29. D. M. Grebenschikov, N. A. Kovrijnikh, and S. A. Kozlov, *Opt. Spectrosc.* **31**, 214 (1971).
30. E. V. Shpol'skii, *Sov. Phys. Usp.* **5**, 522 (1962).
31. V. I. Mikhailenko, Y. R. Redkin, and V. P. Grosul, *Opt. Spectrosc.* **39**, 50 (1975).
32. M. Lamotte, A. M. Merle, J. Joussot-Dubien, and F. Dupuy, *Chem. Phys. Lett.* **35**, 410 (1975).
33. A. M. Merle, W. M. Pitts, and M. A. El-Sayed, *Chem. Phys. Lett.* **54**, 211 (1978).
34. C. Braeuchle, H. Kabza, J. Voitländer, and E. Clar, *Chem. Phys.* **32**, 63 (1978).
35. K. K. Rebane and V. V. Khizhnyakov, *Opt. Spectrosc.* **14**, 193 (1963).
36. J. L. Richards and S. A. Rice, *J. Chem. Phys.* **54**, 2014 (1971).
37. K. K. Rebane, *Impurity Spectra of Solids* (Plenum Press, New York, 1970), pp. 35–124.
38. M. N. Sapozhnikov, *Phys. Status Solidi* **75**, 11 (1976).
39. M. N. Sapozhnikov, *J. Chem. Phys.* **68**, 2352 (1978).
40. G. W. Canters, M. Noort, and J. H. vander Waals, *Chem. Phys. Lett.* **30**, 1 (1975).
41. G. F. Kirkbright and C. G. de Lima, *Analyst* **99**, 338 (1974).
42. B. S. Causey, G. F. Kirkbright, and C. G. de Lima, *Analyst* **101**, 367 (1976).
43. R. Farooq and G. F. Kirkbright, *Analyst* **101**, 566 (1976).
44. J. A. G. Drake, D. W. Jones, B. S. Causey, and G. F. Kirkbright, *Fuel* **57**, 663 (1978).
45. R. I. Personov and T. A. Teplitskaya, *J. Anal. Chem. USSR* **20**, 1176 (1965).

46. L. M. Shabad and G. A. Smirnov, *Atmos. Environ.* **6**, 153 (1972).
47. A. Colmsjö and U. Stenberg, *Anal. Chem.* **51**, 145 (1979).
48. A. Colmsjö and U. Stenberg, *J. Chromatogr.* **169**, 205 (1979).
49. A. P. D'Silva, G. J. Oestreich, and V. A. Fassel, *Anal. Chem.* **48**, 915 (1976).
50. C. S. Woo, A. P. D'Silva, V. A. Fassel, and G. J. Oestreich, *Environ. Sci. Technol.* **12**, 173 (1978).
51. C. S. Woo, A. P. D'Silva, and V. A. Fassel, *Anal. Chem.* **52**, 159 (1980).
52. J. D. Winefordner, *Acc. Chem. Res.* **2**, 361 (1969).
53. R. J. Lukasiewicz and J. D. Winefordner, *Talanta* **19**, 381 (1972).
54. L. A. Mishina and L. A. Nakhimovskaya, *Opt. Spectrosc.* **36**, 298 (1974).
55. N. S. Dokunikhin, V. A. Kizel, M. N. Sapozhnikov, and S. L. Solodar, *Opt. Spectrosc.* **25**, 42 (1968).
56. D. M. Grebenshchikov, N. A. Kovizhynkh, and S. A. Kozlov, *Opt. Spectrosc.* **37**, 155 (1974).
57. G. L. LeBel and J. D. Laposa, *J. Mol. Spectrosc.* **41**, 249 (1972).
58. J. J. Dekkers, G. P. Hoornweg, C. Maclean, and N. H. Velthorst, *J. Mol. Spectrosc.* **68**, 56 (1977).
59. E. V. Shpol'skii, L. A. Klimova, G. N. Nersesova, and V. I. Glyadkovskii, *Opt. Spectrosc.* **24**, 25 (1968).
60. R. C. Stroupe, P. Tokousbalides, R. B. Dickinson, Jr., E. L. Wehry, and G. Mamantov, *Anal. Chem.* **49**, 701 (1977).
61. E. V. Shpol'skii and T. N. Bolotnikova, *Pure Appl. Chem.* **37**, 183 (1974).
62. T. Y. Gaevaya and A. Y. Khesina, *J. Anal. Chem. USSR* **29**, 1913 (1974).
63. R. I. Personov, *J. Anal. Chem. USSR* **17**, 503 (1962).
64. G. E. Fedoseeva and A. Y. Khesina, *J. Appl. Spectrosc. USSR* **9**, 838 (1968).
65. G. E. Kanil'tseva and A. Y. Khesina, *J. Appl. Spectrosc. USSR* **5**, 196 (1966).
66. R. J. McDonald, L. M. Logan, I. G. Ross, and B. K. Selinger, *J. Mol. Spectrosc.* **40**, 137 (1971).
67. R. J. McDonald and B. K. Selinger, *Aust. J. Chem.* **24**, 249 (1971).
68. R. A. Keller and D. E. Breen, *J. Chem. Phys.* **53**, 2562 (1965).
69. A. P. D'Silva, Y. Yang, and V. A. Fassel, "Polynuclear Aromatic Hydrocarbons—Quantitation Using Selectively Excited Time Resolved Luminescence in Shpol'skii Solvents," paper presented at Pittsburgh Conference on Analytical Chemistry and Applied Spectroscopy, Atlantic City, New Jersey, March 13, 1980, Abstract No. 730; Y. Yang, A. P. D'Silva, V. A. Fassel, and M. Iles, *Anal. Chem.* **52**, 1350 (1980); **53**, 894 (1981).
70. E. Whittle, D. A. Dows, and G. C. Pimentel, *J. Chem. Phys.* **22**, 1943 (1954).
71. I. Norman and G. Porter, *Nature* **174**, 508 (1954).
72. E. L. Wehry and G. Mamantov, *Anal. Chem.* **51**, 643A (1979).
73. T. G. Matthews and F. E. Lytle, *Anal. Chem.* **51**, 583 (1979).
74. A. A. Lamola and N. J. Turro, *Energy Transfer and Organic Photochemistry* (Wiley, New York, 1969), p. 37.
75. J. S. Shirk and A. M. Bass, *Anal. Chem.* **41** (11), 103A (1969).
76. R. C. Stroupe, P. Tokousbalides, R. B. Dickinson, Jr., E. L. Wehry, and G. Mamantov, *Anal. Chem.* **49**, 701 (1977).
77. B. Meyer, *Science* **168**, 783 (1970).
78. A. M. Bass and H. P. Broida, *Formation and Trapping of Free Radicals* (Academic Press, New York, 1960).
79. H. E. Hallam, *Vibrational Spectroscopy of Trapped Species* (Wiley, London, 1973).
80. S. Cradock and A. J. Hinchcliffe, *Matrix Isolation: A Technique for the Study of Reactive Inorganic Species* (Cambridge University Press, New York, 1975).

81. M. Moskovits and G. A. Ozin, *Cryochemistry* (Wiley, New York, 1976).
82. Ref. 1, p. 229.
83. D. M. Hembree, E. R. Hinton, Jr., R. R. Kemmerer, G. Mamantov, and E. L. Wehry, *Appl. Spectrosc.* **33**, 477 (1979).
84. P. Tokousbalides, E. R. Hinton, Jr., R. B. Dickinson, Jr., P. V. Bilotta, E. L. Wehry, and G. Mamantov, *Anal. Chem.* **50**, 1189 (1978).
85. J. R. Maple, E. L. Wehry, and G. Mamantov, *Anal. Chem.* **52**, 920 (1980).
86. P. Tokousbalides, E. L. Wehry, and G. Mamantov, *J. Phys. Chem.* **81**, 1769 (1977).
87. S. Cova and A. Longoni, in *Analytical Laser Spectroscopy*, N. Omenetto, ed. (Wiley, New York, 1979), p. 484.
88. E. L. Wehry, J. R. Maple, R. R. Gore, and R. B. Dickinson, Jr., paper presented at the 180th National American Chemical Society Meeting, Las Vegas, Nevada, August 1980.
89. T. Hirschfeld, *Appl. Spectrosc.* **31**, 245 (1977).
90. R. B. Dickinson, Jr., and E. L. Wehry, *Anal. Chem.* **51**, 778 (1979).
91. T. Hirschfeld, *Anal. Chem.* **52**, 297A (1980).
92. E. L. Wehry, G. Mamantov, D. M. Hembree, and J. R. Maple, in *Polynuclear Aromatic Hydrocarbons: Chemistry and Biological Effects*, A. Bjorseth and A. J. Dennis, eds. (Battelle Press, Columbus, Ohio, 1980), p. 1005.
93. G. T. Reedy, S. Bourne, and P. T. Cunningham, *Anal. Chem.* **51**, 1535 (1979).
94. S. Bourne, G. T. Reedy, and P. T. Cunningham, *J. Chromatogr. Sci.* **17**, 460 (1979).
95. A. Szabo, *Phys. Rev. Lett.* **25**, 924 (1970).
96. A. Szabo, *Phys. Rev. Lett.* **27**, 323 (1971).
97. R. I. Personov, E. I. Al'shits, and L. A. Bykovskaya, *JETP Lett.* **15**, 431 (1972).
98. R. I. Personov, E. I. Al'shits, L. A. Bykovskaya, and B. M. Kharlamov, *Sov. Phys. JETP* **38**, 912 (1974).
99. L. A. Bykovskaya, R. I. Personov, and B. M. Kharlamov, *Chem. Phys. Lett.* **27**, 80 (1974).
100. R. I. Personov and E. I. Al'shits, *Chem. Phys. Lett.* **33**, 85 (1975).
101. E. I. Al'shits, R. I. Personov, A. M. Pyndyk, and V. I. Stogov, *Opt. Spectrosc.* **39**, 156 (1975).
102. J. H. Eberly, W. C. McColgin, K. Kawaoka, and A. P. Marchetti, *Nature* **251**, 215 (1974).
103. A. P. Marchetti, W. C. McColgin, and J. H. Eberly, *Phys. Rev. Lett.* **35**, 387 (1975).
104. A. P. Marchetti, M. Scozzafava, and R. H. Young, *Chem. Phys. Lett.* **51**, 424 (1977).
105. W. C. McColgin, A. P. Marchetti, and J. H. Eberly, *J. Am. Chem. Soc.* **100**, 5622 (1978).
106. I. Abram, R. A. Auerbach, R. R. Birge, B. E. Kohler, and J. M. Stevenson, *J. Chem. Phys.* **61**, 3857 (1974).
107. I. I. Abram, R. A. Auerbach, R. R. Birge, B. E. Kohler, and J. M. Stevenson, *J. Chem. Phys.* **62**, 2473 (1975).
108. G. Flatscher, K. Fritz, and J. Friedrich, *Z. Naturforsch.* **31a**, 1220 (1976).
109. G. Flatscher and J. Friedrich, *Chem. Phys. Lett.* **50**, 32 (1977).
110. K. Cunningham, J. M. Morris, J. Fünfschilling, and D. F. Williams, *Chem. Phys. Lett.* **32**, 581 (1975).
111. J. Fünfschilling and D. F. Williams, *Appl. Spectrosc.* **30**, 443 (1976).
112. J. Fünfschilling and D. F. Williams, *Photochem. Photobiol.* **26**, 109 (1977).
113. J. M. Hayes and G. J. Small, *Chem. Phys. Lett.* **54**, 435 (1978).
114. J. M. Hayes and G. J. Small, *Chem. Phys. Lett.* **27**, 151 (1978).
115. M. R. Topp and H. Lin, *Chem. Phys. Lett.* **50**, 412 (1977).
116. T. E. Orlowski and A. H. Zewail, *Chem. Phys. Lett.* **70**, 1390 (1979).
117. J. P. Lemaistre and A. H. Zewail, *Chem. Phys. Lett.* **68**, 302 (1979).
118. R. I. Personov, E. I. Al'shits, and L. A. Bykovskaya, *Opt. Commun.* **6**, 169 (1972).
119. R. I. Personov and B. M. Khablamovanov, *Opt. Commun.* **7**, 417 (1973).

120. M. Labhart, G. Miklas, J. Keller, and U. P. Wild, *J. Chem. Phys.* **72**, 1764 (1980).

121. J. C. Brown, M. C. Edelson, and G. J. Small, *Anal. Chem.* **50**, 1394 (1978).

122. E. L. Wehry, in *Practical Fluorescence: Theory, Methods, and Techniques*, G. G. Guilbault, ed. (Marcel Dekker, New York, 1973), pp. 124–126.

123. P. Avouris, A. Campion, and M. A. El-Sayed, *Chem. Phys. Lett.* **50**, 9 (1977).

124. R. R. Gore and E. L. Wehry, unpublished results, University of Tennessee, 1980.

125. E. I. Al'shits, R. I. Personov, and B. M. Kharlamov, *Chem. Phys. Lett.* **40**, 116 (1976).

126. E. I. Al'shits, R. I. Personov, and B. M. Kharlamov, *Opt. Spectrosc.* **41**, 474 (1976).

127. T. B. Tamm and P. M. Saari, *Chem. Phys. Lett.* **30**, 129 (1975).

128. R. L. Williamson and A. L. Kwiram, *J. Phys. Chem.* **83**, 3393 (1979).

129. J. Fünfschilling, E. Wasmer, and I. Zschokke-Granacher, *J. Chem. Phys.* **69**, 2949 (1978).

130. H. Kubota, W. H. Griest, and M. R. Guerin, in *Trace Substances in Environmental Health*, D. D. Hemphill, ed. (University of Missouri Press, Columbia, Missouri, 1975), p. 281.

131. B. M. Kharlamov, R. I. Personov, and L. A. Bykovskaya, *Opt. Commun.* **12**, 191 (1974).

132. B. M. Kharlamov, L. A. Bykovskaya, and R. I. Personov, *Opt. Spectrosc.* **42**, 445 (1977).

133. J. C. Brown, J. A. Duncanson, Jr., and G. J. Small, *Anal. Chem.* **52**, 1711 (1980).

134. D. H. Levy, L. Wharton, and R. E. Smalley, in *Chemical and Biochemical Applications of Lasers*, C. B. Moore, ed. (Academic Press, New York, 1977), p. 1.

135. R. E. Smalley, L. Wharton, D. H. Levy, and D. W. Chandler, *J. Chem. Phys.* **68**, 2487 (1978).

136. A. Amirav, U. Even, and J. Jortner, *Opt. Commun.* **32**, 266 (1980).

137. J. C. Brown, J. M. Hayes, and G. J. Small, "New Laser Based Methodologies for the Determination of Organic Pollutants by Fluorescence," paper presented at 32nd Summer Symposium, ACS Division of Analytical Chemistry, West Lafayette, Indiana, June 28, 1979.

Chapter 7

Use of Luminescence Spectroscopy in Oil Identification

DeLyle Eastwood

A. INTRODUCTION

Recent advances in fluorescence/luminescence spectroscopy and related instrumentation have resulted in effective methods for forensic oil spill identification. This chapter will summarize recent research on molecular emission methods for oil identification as well as related fluorescence studies on petroleum concerning, for example, the effects of weathering, pattern recognition, detection and quantitation at low levels in water, and both on-site and remote-sensing applications.

The main advantage of fluorescence techniques for oil identification is their high sensitivity to the aromatic components of petroleum oil as well as their potentially great specificity or selectivity when the correct technique or combination of techniques is utilized. Fluorescence also has the capability for oil identification directly in water or in a thin oil film without the need for elaborate sample preparation. Thus, fluorescence can also be utilized in flow-through or remote-sensing applications.

Where known oil samples or comparable oil reference fluorescence standards are available, rapid *in situ* quantitation becomes feasible. For low concentrations in water, where the oil becomes fractionated by differential solubilization and evaporation processes, the distinction between petroleum oil and aromatic hydrocarbon analysis becomes unclear.

DELYLE EASTWOOD • Chemistry Branch, U.S. Coast Guard Research and Development Center, Avery Point, Groton, Connecticut 06340

It should be clearly understood that the forensic application of oil identification is that of matching spectral "fingerprints" of spilled oils with suspects, rather than trying to identify specific chemicals in the oil. The main problems in fluorescence oil identification have been in correcting for weathering and in developing suitably quantitative pattern recognition or matching techniques for the relatively broad or structureless oil fluorescence spectra. These difficulties will be shown to have now been largely eliminated.

B. INSTRUMENTAL AND COMPUTER TECHNIQUES AT ROOM AND LOW TEMPERATURES

1. Excitation and Emission Spectroscopy

Parker[1] and others[2–18] have shown that fluorescence emission spectroscopy provides a satisfactory oil identification method, with an approximately 250-nm excitation yielding the most structure for petroleum oils in general. Such emission spectra at room temperature can be satisfactorily generated on any commercially available spectrofluorometer that has adequate resolution (2–5 nm).

Both the American Society of Testing and Materials (ASTM) method D3650-78[19] and the similar U.S. Coast Guard (USCG) fluorescence method (which is part of the Coast Guard Oil Spill Identification System[20] multimethod approach) utilize fluorescence emission with excitation at 254 nm and at an oil concentration of 20 ppm in spectroquality cyclohexane. Figures 1 and 2 show the room- and low-temperature spectra for a light Australian crude oil and a marine diesel lubricating oil, respectively, with excitation at 254 nm. Excitation at 254 nm permits the use of Hg as well as Xe lamps and provides a wavelength at which many aromatic compounds absorb. The 20-ppm oil concentration for this method represents a compromise such that the method will be applicable for all oil types. Additional emission spectra are sometimes also recorded at other excitation wavelengths, such as 290, 330, or 340 nm, for additional information, but they generally yield less spectral structure. Figure 3 indicates the effect of varying the excitation wavelength on the spectra of a heavy crude oil from Venezuela.

Fluorescence excitation spectra measured with detection at the principal emission peaks are sometimes useful for oils with relatively little spectral structure in the emission spectra. This procedure, however, is not specified in the ASTM or USCG methods, since principal wavelengths for detection of excitation spectra cannot be determined in advance but are chosen from the emission spectra.

Although corrected excitation and emission spectra are not necessary when all the spectra to be compared are recorded with the same spectro-

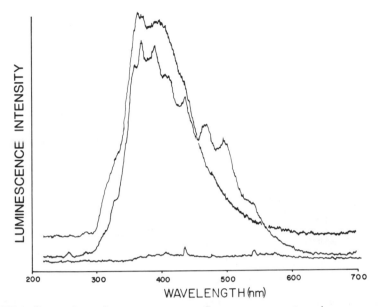

FIGURE 1. Comparison of room-temperature fluorescence spectrum (upper curve) with low-temperature luminescence spectrum (lower curve) for a light Australian crude oil (10 ppm in methylcyclohexane). Excitation wavelength is 254 nm; excitation bandwidth, 6 nm; emission bandwidth 4 nm. Instrumentation was the Baird Fluorispcc SF-100 spectrofluorometer (uncorrected). Taken from an oral presentation by Fortier and Eastwood.[66]

fluorometer, the capability to generate corrected spectra is desirable, especially when different instruments or interlaboratory comparisons are involved.

2. Synchronous and Variable Synchronous Spectroscopy*

Lloyd[21–24] described the advantages of synchronous excitation techniques for forensic oil identification, especially for automotive engine oils. In this tcchnique, both excitation and emission monochromators are scanned simultaneously at the same speed with a fixed wavelength separation, or offset, between them. An offset of 25 nm, which yielded the most spectral structure for oils, is comparable to the typical separation between absorption and emission maxima for aromatic hydrocarbons. Spectra corresponding to offsets such as 15 and 40 nm should also be measured for closely similar oils. Vo-Dinh et al.[25,26] have recently studied aromatic mixtures and

---

* A detailed discussion of synchronous luminescence techniques is provided by Vo-Dinh in Chapter 5 of the present volume.

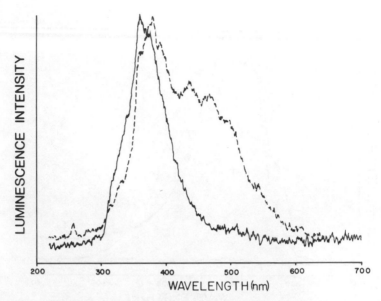

FIGURE 2. Comparison of the room-temperature fluorescence spectrum (——) with the low-temperature luminescence spectrum (– – –) of a marine diesel lubricating oil with operating conditions as in Figure 1.

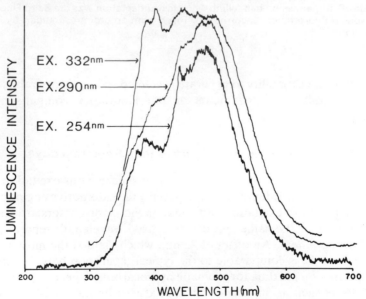

FIGURE 3. Comparison of low-temperature luminescence spectra of a heavy Venezuelan crude oil at three different excitation wavelengths: 254, 290, and 332 nm (with operating conditions as in Figure 1).

synthoils using much smaller offsets (less than 5 nm) than those advocated by Lloyd.

In synchronous excitation spectroscopy, there is an approximate separation of fluorescence peaks with wavelength according to the number of rings in the aromatic system, since for a homologous series the longer aromatic ring systems tend to absorb and emit at longer wavelengths. Thus, greater spectral structure is obtained by this technique, especially for heavier oils such as No. 6 fuel oils, which usually exhibit relatively featureless conventional emission spectra. Recent studies[27-31] have indicated that synchronous excitation spectra may exhibit greater changes with weathering than conventional emission spectra and therefore such weathering effects should be further investigated before this technique is utilized with weathered oils. Figure 4 shows uncorrected low-temperature luminescence synchronous spectra for a heavy crude oil from Ecuador, one sample unweathered and the other weathered 1 week.

Very recently, Kubic et al.[5,32,33] demonstrated the potential of variable synchronous excitation spectroscopy for the identification of lubricating oils. In this technique, emission and excitation monochromators are scanned simultaneously but at different rates, with the emission monochromator

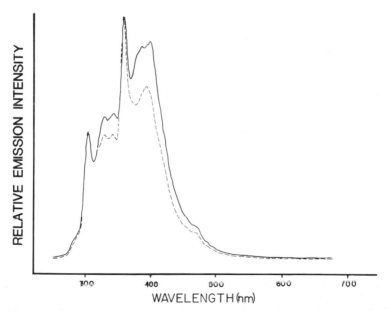

FIGURE 4. Low-temperature luminescence synchronous excitation spectra (wavelength offset = 25 nm) for a heavy Ecuadorian crude oil. Neat (——) and weathered 1 week (– – –), with operating conditions as in Figure 1.

always scanning faster. Kubic observed that particular ratios of scanning speeds for the two monochromators and relative offsets could often be found that provided greater spectral structure for variable synchronous excitation spectra than for either conventional emission or synchronous excitation spectra of the same oils.

3. Difference Spectroscopy

Difference spectrofluorometry for oil identification has been demonstrated by several chemists[34,35] since instrumentation capable of generating real-time optical difference spectra became recently commercially available. The difference technique has the disadvantage of decreased sensitivity as compared to the standard fluorescence emission method; however, it has the capability of eliminating solvent Raman peaks and background fluorescence and, in some cases, of correcting for sample contamination. The main advantage of this approach is that minor spectral differences are enhanced, which renders difference spectroscopy useful for direct spectral comparison of closely similar oils or for observations of spectral differences caused by weathering. Figure 5 shows a typical example of a difference spectrum. With digitized data, a computer can also generate difference spectra from

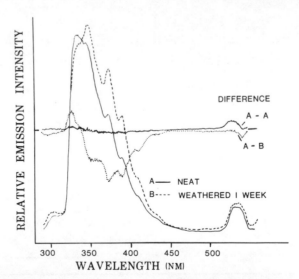

FIGURE 5. Room-temperature emission and difference spectra for a No. 2 fuel oil (20 ppm in methylcyclohexane). Excitation wavelength, 254 nm; excitation bandwidth, 10 nm; emission, 2 nm. Spectrum A represents a neat (unweathered) No. 2 fuel oil; spectrum B represents the same oil weathered 1 week. Difference spectrum A–A is a measure of error due to background and the difference in cells used, etc.; spectrum A–B is the difference spectrum due to weathering. Instrumentation was the Farrand Mark I spectrofluorometer (corrected). Taken from an oral presentation by Eastwood *et al.*[84]

normalized, sequentially recorded spectra; however, this approach lacks the advantages of real-time comparison and may increase errors due to instrumental variations.

4. Derivative Spectroscopy

Derivative spectroscopy as described by O'Haver[36,37] and others[38-40] has also been used to enhance minor spectral features in oil fluorescence spectra. In this technique, the first or higher derivative of the fluorescence emission is plotted against wavelength using either electronic- or computer-generated data. Although this technique is especially promising for visual matching of oil spectra, it has not yet been extensively tested for this application. Commercially available microprocessor recorders and spectrofluorometer accessories have recently made this technique generally accessible so that it will probably be more widely used. Figure 6 shows an emission spectrum for a No. 6 fuel oil with spectra of the first and second derivatives.

5. Contour (Total Luminescence) Spectroscopy

The most universal fluorescence/luminescence method is the contour or "total luminescence" method in which the excitation and emission wavelengths for oil emission are used as axes for plotting contours of equal fluorescence intensities.* The contour spectrum truly constitutes the most complete fluorescence spectral fingerprint of each petroleum oil. Recently, interest in the use of this technique for oil identification has developed from the work of Hornig et al.,[41-43] Warner et al.,[44,45] and others.[46-51] This method has great potential for oil and complex aromatic hydrocarbon mixtures, particularly as a means for determining optimum conditions for measuring oils by simpler fluorescence methods.

Until recently, contour spectra were relatively slow to generate without expensive custom-made equipment, but this difficulty has been overcome with the advent of computerized commercial equipment now available. Figures 7 and 8 show typical contour spectra for a No. 6 fuel oil at room and low temperatures, respectively. These complex spectra have been relatively difficult to interpret; however, methods to quantitate their informational content are now being developed.[52]

6. Special Low-Temperature Techniques†

Low-temperature luminescence results in increased spectral structure for oil identification as indicated in several previous studies.[53-56] The Coast

---

* This topic is discussed in more detail in Chapter 4 of the present volume.
† See Chapter 6 for a detailed discussion of low-temperature techniques in analytical fluorometry.

EMISSION SPECTRA NO. 6 OILS

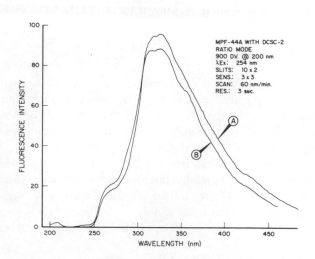

DERIVATIVE SPECTRA NO. 6 OILS

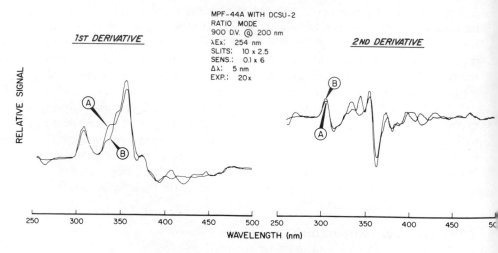

FIGURE 6. Comparison of room-temperature emission spectra with first- and second-derivative spectra for two similar No. 6 fuel oils. Excitation wavelength, 254 nm; excitation bandwidth, 10 nm; emission, 2.5 nm. Instrumentation was Perkin–Elmer MPF-44A spectrofluorometer used in the ratio mode. Courtesy of J. L. DiCesare and T. J. Porro[38]; reprinted with permission.

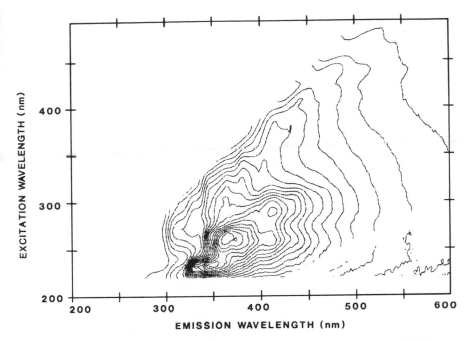

FIGURE 7. Room-temperature fluorescence contour spectrum of a No. 6 fuel oil (20 ppm in methylcyclohexane). Excitation and emission bandwidths are both 5 nm. Instrumentation was Baird Total Luminescence Spectroscopy System (TLS 1000). Courtesy of L. Glering and D. Busch; and was previously used in oral presentation by Eastwood and Hendrick.[4]

Guard R & D Center[20,57] has employed low-temperature luminescence for oil identification as a backup technique to distinguish among closely similar oils. Figure 9 shows an example of oil fingerprinting by low-temperature luminescence for a real-world spill case involving No. 2 fuel oils. All of the previously discussed techniques and combinations thereof can be used at low temperatures (usually 77°K) employing special high-quality solvents, e.g., methylcyclohexane, for glass formation. Low-temperature luminescence has the advantage of increasing spectral structure and sensitivity for the peaks due to fluorescence, as well as providing additional spectral peaks due to phosphorescence. Operating requirements such as quartz or Suprasil® Dewars and sample tubes and the need to minimize scatter have led to the avoidance of low-temperature luminescence methods, although in practice these problems can easily be addressed. Unfortunately, not all fluorescence instrumentation is equally effective in reducing scatter or in eliminating background fluorescence, so spectrofluorometers and accessories should be carefully screened for this application.

A variety of low-temperature methods whose potentials have not yet been adequately explored will be summarized.

Time-resolved phosphorimetry has been explored for oils,[38] using both mechanical phosphoroscopes (down to 0.5 msec) and pulsed Xe sources (down to 10 μsec). The only significant variation with time delay noted in the time-resolved phosphorescence oil spectra was the enhancement of the vanadyl petroporphyrin emission at approximately 710 nm when using delay times as short as 20 μsec.[58] Vanadyl etioporphyrin was earlier found by Eastwood and Gouterman[59] to have a phosphorescence lifetime of 60 μsec. The lack of variation with time delay is perhaps not too surprising, since many of the aromatic components in oils have lifetimes in the range of seconds.

There is usually a degree of spectral overlap between the fluorescence and phosphorescence components in the low-temperature luminescence of an oil sample. Preliminary investigations[60,61] have indicated that phosphorescence spectra, although often less structured then low-temperature

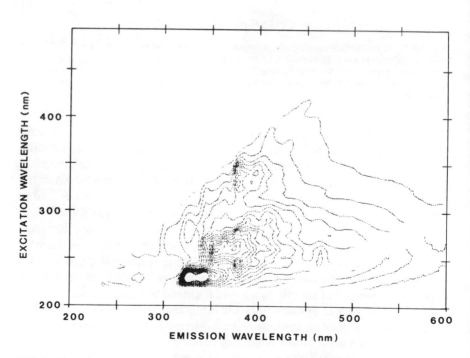

FIGURE 8. Low-temperature luminescence contour spectrum of a No. 6 fuel oil (operating conditions and credits as in Figure 7).

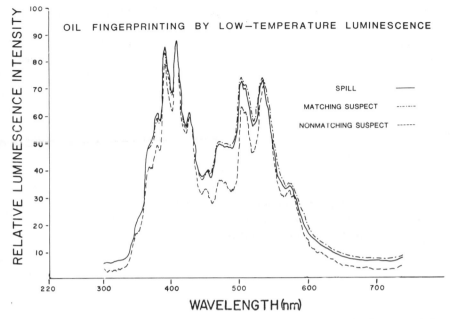

FIGURE 9. Example of the low-temperature luminescence method as used for identification purposes in a real-world spill case. Three No. 2 fuel oils are compared: the spill sample (——); the suspect sample that was a definite match (– · –); and the suspect sample that was a definite nonmatch (– – –). From Fortier and Eastwood[57]; reprinted with permission from the American Chemical Society.

fluorescence spectra, are also useful for oil identification. The phosphorescence spectral structure may be enhanced by use of a solvent containing heavy atoms such as 1% bromoform in methylcyclohexane. Perfluoro(1,3-dimethylcyclohexane) was also found[58] to be a useful organic glass for sharpening and intensifying the phosphorescence for oil samples. Figure 10 gives an example of the use of this glass for low-temperature polarized emission spectra of a No. 2 fuel oil.

Preliminary low-temperature studies of polarized emission and excitation spectra of different oil classes indicated that reproducible polarization differences among oils occur and furnish another parameter for oil identification. Figure 11 shows the four polarized emission spectra for an Alaskan crude oil, while Figure 12 shows polarized excitation spectra for a No. 2 fuel oil and the corresponding polarization curve. An important potential source of error in polarization measurements is the variability in polarization due to strains in the organic glass, but this can be minimized by cooling the samples slowly and reproducibly or by judicious selection of relatively strain-free glasses.

Temperature-variation measurements are another possible method of optimizing techniques for oil identification by low-temperature luminescence. Figure 13 shows results of a preliminary temperature-variation study on a No. 4 fuel oil (temperatures were only approximate).

Several investigators have used very low temperatures and/or either special Shpol'skii solvents such as octane[62-65] or matrix isolation techniques[66-69] to get very sharp quasi-line structured spectra. The use of lasers permitting matrix site-selective excitation can yield relatively sharp spectra even at 77°K. Removal of polar fractions of the oil can also enhance the observed spectral structure. Degassing techniques have also been reported to enhance luminescence structure and intensity.[56] These special techniques allow some polynuclear aromatic hydrocarbon components to be identified and quantified in the presence of complex oil samples.

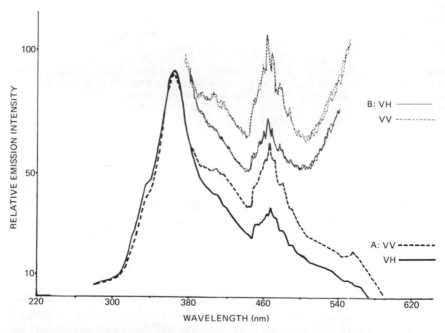

FIGURE 10. Example of the use of perfluoro(1,3-dimethylcyclohexane) as an organic glass to enhance the structure of phosphorescence spectra at 77°K, showing polarizer emission spectra of a No. 2 fuel oil (at a concentration of 20 ppm). Instrumentation was the Farrand Mark I spectrofluorometer. Excitation wavelength, 254 nm; excitation bandwidth, 10 nm; emission bandwidth 1 nm (upper spectra); emission bandwidth 2.5 nm (lower spectra). Polarizer orientations VH (——) and VV (– – –). Taken from oral presentation by Eastwood and Hendrick.[58]

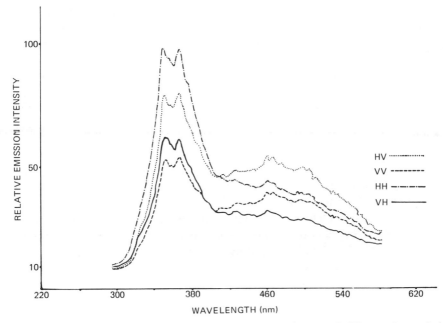

FIGURE 11. Polarized emission spectra of an Alaskan crude oil (20 ppm in methyl-cyclohexane) at 77°K. Excitation wavelength, 254 nm; excitation bandwidth, 10 nm; emission, 2.5 nm. Polarizer orientations are HV (· · ·), VV (– – –), HH (– · –), and VH (———). Instrumentation was the Farrand Mark I spectrofluorometer (corrected). Taken from an oral presentation by Eastwood and Hendrick.[58]

7. Lifetime Measurements

Fluorescence lifetimes have been shown to vary for different classes of oils[70], thereby providing an additional parameter for identification or classification. This technique, which should be especially useful for remote sensing, has not yet been extensively pursued. Two techniques are available for measuring lifetimes; the pulsed-source technique and the phase-shift technique.* Although the latter technique has been demonstrated in the past mainly for aromatic hydrocarbons, its potential for oil identification has been shown by recent unpublished work.[71]

8. Miscellaneous Chromatographic Techniques

Oil identification by combining various chromatographic techniques with fluorescence examination of the various separated oil fractions has

---

* A more detailed discussion of the techniques for measuring fluorescence decay times is presented in Chapter 2 of the present volume.

been explored by a number of authors including Drushel and Sommers,[72] McKay and Latham[73] and Jadamec, Saner, and Talmi.[74 76] In early work, Drushel and Sommers[72] combined gas chromatography with fluorescence emission, excitation, and phosphorescence spectra to examine petroleum fractions containing aromatic sulfur and nitrogen compounds. Brownrigg and Hornig[77] also investigated the identification of oil components by gas chromatography combined with low-temperature luminescence. McKay and Latham[13] combined gel permeation chromatography with fluorescence to identify important polynuclear aromatic hydrocarbon components in high-boiling petroleum distillates.

Others have recently explored oil identification combining high-pressure liquid chromatography with sophisticated optoelectronic image detectors such as videofluorometers[39,45,78] and optical multichannel analyzers (OMA).[74-76,79] Such techniques have tremendous potential in that they can counter many effects due to weathering and contamination. Such instrumentation interfaced with computers can rapidly generate emission/excitation spectra of each eluted fraction, and with appropriate pattern recog-

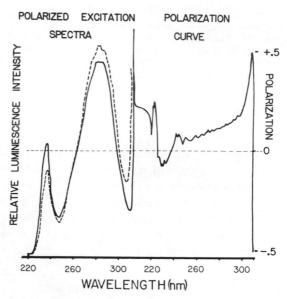

FIGURE 12. Polarized excitation spectra and polarization curve for a No. 2 fuel oil (20 ppm in methylcyclohexane) at 77°K. Detection wavelength, 320 nm; excitation bandwidth, 2.5 nm; emission bandwidth 20 nm. Polarization orientations are VV (——) and VH (– – –). Instrumentation was the Perkin–Elmer MPF-44A spectrofluorometer. Courtesy of D. Terhaar and T. Porro; previously used in oral presentation by Eastwood and Hendrick.[58]

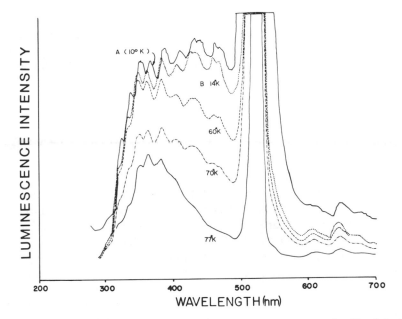

FIGURE 13. Temperature-variation study of luminescence spectra of a No. 4 fuel oil (20 ppm in methylcyclohexane) over the temperature range of 77–10°K (temperatures approximate). Excitation wavelength, 254 nm; excitation bandwidth, 14 nm; emission bandwidth 2 nm (curve A) and 4 nm (curve B and all remaining curves). Instrumentation was Baird Fluorispec SF-100 spectrofluorometer (uncorrected). Taken from oral presentation by Eastwood and Hendrick.[58]

nition programs they could be used for oil identification. The disadvantages of this approach are the expensive and sophisticated equipment and the tremendous quantity of data to be processed for each oil sample.

C. WEATHERING

Most studies on oil identification by fluorescence and low-temperature luminescence have dealt almost exclusively with unweathered oils. Obviously, the influence of oil weathering on the success of an analytical technique such as fluorescence for oil identification is an important concern. A number of studies[80–83] have indicated that the primary mechanisms of short-term weathering are solubilization, evaporation, and emulsification. These weathering processes, affected in turn by factors such as air and water temperatures, wind speed, and wave motion, are the most important considerations in the initial weathering period of up to 3 days. Other factors, such as biodegradation (particularly noticeable in tar balls) and photo-oxidation, are not major causes of oil alteration in short-term weathering. Initial studies[84] have indicated that oils can be identified by room-temperature fluorescence or low-temperature luminescence after normal weathering for periods of 1 week for heavy oils or up to 1 or 2 days for light oils. These weathering periods produce a spectral change of less than 4% in major peak amplitudes for heavy oils, or as much as 10% for some light oils.

Artificial weathering of laboratory oil samples coupled with special computer analysis techniques allows for more quantitative determination over longer weathering periods.[85]

Except for emission spectra at room and low temperatures, other fluorescence/luminescence techniques have not yet been adequately explored on a sufficiently large number of weathered oil samples to determine those spectral features that are most persistent with weathering.

Another way to circumvent weathering effects is to combine high-pressure liquid chromatography with fluorescence detection, e.g., with optical multichannel analyzer detection.[74–76,79] Such a fractionation separates various oil components and the lighter-weight and more polar compounds most affected by weathering can be ignored. Fractional distillation of oil samples to remove light ends from the oils is also an effective technique for minimizing the effects of weathering.

A combination of techniques including artificial weathering, computer simulation, and, if necessary, chromatographic fractionation will permit fluorescence identification to be extended to longer weathering periods.

D. PATTERN RECOGNITION AND STATISTICAL ANALYSIS

Until recently, very little attention has been directed towards pattern recognition and statistical analysis of fluorescence spectra for oil identification. These areas are essential to a quantitative approach for oil identification, which is especially important for forensic purposes. The first approach to pattern recognition of fluorescence spectra was developed by Curtis,[86] who divided 230 fluorescence spectra of various oil types into 33 groups, or clusters. These clusters were then categorized in terms of "peaks" and "valleys," utilizing in addition to their wavelengths and relative intensities a factor called "sharpness." Starks and Curtis[87] also used an approach based on distance measures and cluster analysis to prove the independence of fluorescence and infrared spectral data taken on the same oils.

Chien and Killeen[88–89] also have addressed pattern recognition techniques as applied to fluorescence spectra of oils and reaffirmed the statistical independence of fluorescence and infrared spectral data. In this approach, because of the relatively broad fluorescence spectra for oils, rather than using discrete peaks as has been done for infrared pattern recognition, a continuously digitized (every 1 or 2 nm) set of 100–200 data points per spectrum was generated. A Bayer probabilistic model for matching oil spills with suspects was also developed by Killeen and Chien.[90]

Warner et al.[44] have developed a package of computer algorithms that can characterize multicomponent fluorescence samples.* Requiring for input data the emission–excitation matrix whose elements are the fluorescent intensity values of the emission–excitation spectra, the algorithms can:

(a) determine a lower bound on the number of spectral components that are present in the sample, or
(b) determine the amount of each component when its presence is known, or
(c) approximate the emission and excitation spectra for unknown components.

These algorithms utilize the mathematical techniques of eigenvector analysis, least squares, and linear programming. Successful results have been obtained for simulated data; problems of real-world oil samples are still being addressed.

Recently Killeen et al.[91] developed a vector analysis method for identifying weathered oils. In this approach the digitized fluorescence spectra are considered as n-dimensional vectors (n = number of digitized points in each spectrum) and the angular distance between vectors in n space is used as a distance measure for differences in such spectra. Spectra of weathered spilled samples are then compared with spectra of suspect oils by approximating the weathering surface for each suspect oil. This is done by means of the hyperplane generated by the vector of the neat suspect oil spectrum and that corresponding to the laboratory-weathered suspect oil sample. Then the angle between the spill spectrum and its projection on the hyperplane proves to be an excellent quantitative measure of the likelihood of a match or a nonmatch. This technique has been shown to be valid for interpolation or extrapolation over short weathering periods (up to 1 week). Although this vector analysis method has so far been tested on only a few weathered oil samples, this procedure appears to be more generally applicable.

The possible advantages of "total luminescence" computerized spectra for pattern recognition of oils have been pointed out in several publications by Hornig, Giering, and Brownrigg.[41–43,46–50,92] This technique should be

---

* This subject is considered in detail in Chapter 4 of the present volume.

particularly useful for oil classification or for difference contours to examine spectral changes due to weathering A more quantitative approach to reducing the data contained in contour spectra is presently being developed.[52]

E. DETECTION AND QUANTITATION AT LOW OIL CONCENTRATIONS IN WATER

Fluorescence spectroscopy has been widely employed for monitoring petrogenic aromatic hydrocarbons in water because of its high sensitivity and capability for rapid analysis. The main problem occurs in the selection of appropriate petroleum oils as calibration standards. If an oil of similar class or fluorescence signature is used to calibrate the instrumentation or if a multiwavelength approach compensates for the difference in emission intensities for different classes of oils, concentrations over the range of 1–1000 ppm can be measured to 5% accuracy. Above a few ppm, emulsification into small oil droplets, usually by an ultrasonic technique, must be employed to avoid the possibility of large droplets that would give erroneous readings. At lower concentrations, preconcentration in a suitable solvent such as hexane is required, preferably followed by a separation step to remove naturally occurring fluorescent biogenic polar materials.

Much of the early work on oil quantitation in sea water was done by Gordon and Keizer,[93–96] who excited the hexane extract of the seawater sample at 374 nm (the wavelength at which fluorescence occurs from three- and four-ring polycyclic aromatic hydrocarbons such as anthracene, phenanthrene, and chrysene). They reported average concentrations in surface waters of the Atlantic Ocean of 20.4 μg/liter or within a range from 18–0.4 μg/liter in water at depths from 1 to 5 m. This method did not completely differentiate against gelbstoff (polar biogenically derived material), which also has some emission at 374 nm. Below 0.2 μg/ml oil in water, or more for some coastal or polluted areas, such gelbstoff or non-petroleum pollutants such as lignin sulfonates may become a serious source of error. More recently, Hoffman et al.[97] have made hexane extracts of seawater for fluorescence synchronous excitation techniques ($\Delta\lambda = 25$ nm). This quantitative method employed a suspect oil as a fluorescence standard to establish hydrocarbon concentrations.

In summary, the important problems (largely beyond the scope of this chapter) for all such quantitative studies are the following: the introduction of fluorescent impurities during sampling, distinguishing between petrogenic and biogenic fluorescing materials, the loss of petrogenic material in seawater due to fractionation and chemical reaction, distinguishing between dissolved and emulsified oil and oil adsorbed on particulates, and choosing an appropriate standard or standards when the exact source of the oil is unknown. These sources of error have probably resulted in erroneously elevated published values for petroleum hydrocarbons in seawater. Further, these errors have thus far prevented agreement on a standard method for fluorescence quantitation of oil in water. Further studies to eliminate disagreement on these sources of error are required.

F. REMOTE SENSING AND ON-SITE APPLICATIONS

Fluorescence has obvious applications for remote sensing. The U.S. Coast Guard, EPA, and NASA, as well as various research groups outside of the United States, have carried out investigations involving fluorescence, usually laser excited, as a means of remote sensing of oil contamination. Several review articles[93-101] have recently surveyed the use of fluorescence for the remote detection of oil spills. An ultraviolet laser source and a fluorosensor with a spectral resolution of about 20 nm should offer the greatest potential for oil identification.

A Japanese team, Sato et al.,[102] examined the fluorescence and laser Raman signals obtained remotely from different types of oil using an argon-ion laser (514.5 nm) for excitation. The cw laser mandated nighttime use only; a pulsed source would be required for daytime operation. Earlier, a group from the Canadian Department of Interior[103] had used a cw He–Cd laser at 444.8 nm for the remote sensing of oil.

Work performed by Fantasia et al.[70] employed a pulsed nitrogen laser for oil detection and included fluorescence lifetime measurements. Measures[104-106] also recommended the pulsed nitrogen laser (337 nm) for oil detection because of its high peak power (1 MW), its short duration (2 nsec), and its high repetition rate (100 pulses per second).

Sandness and Ailes[107] used xenon lamps as sources for an active aerial scanner for nighttime detection of water pollutants including oil. Their instrumentation was an imaging system and there was no attempt to classify the pollutants. Bandpass characteristics of the filters limited the detection range from 400 to 600 nm divided into four equal spectral regions. This spectral range permitted only marginal coverage of the fluorescence region for light oils at the long-wavelength end of their emission spectra.

Laboratory studies related to remote sensing of oils have included a study of fluorescence excitation/emission spectra of oil films, varying the thickness and oil type.[108] In addition, a study of marine luminescence spectra[109] provided characteristic fluorescence signatures for seawater background samples as well as spectra of likely water pollutants including oil.

Gross[110] carried out a preliminary field investigation of oil and back-ground luminescence signatures in Los Angeles Harbor. Watson[111] tested a Fraunhofer line discriminator using the Fraunhofer wavelengths at 518.4 or 486.1 nm for remote sensing of luminescing environmental pollutants including oil seeps in the Santa Barbara, California, region.

Eldering et al.[112] developed an oil spill surveillance system using Xe lamps and capable of oil spill detection and oil classification in a harbor environment at a range in excess of 100 m during hours of darkness. This included an estimation of film thickness from remote fluorescence measurements and a polarization technique for discriminating between oil and marine fluorescence. Gram[113-117] developed a fluorescence oil detector designed to be used on a buoy.

One recent prototype oil-in-water monitor developed by Hornig et al.[118-121] uses dual polychromators to present a total luminescence image. A computer-designed mask sums fluorescence from many spectral regions to produce approximately equal responses from many oil mixtures. This monitor responds linearly to oil concentrations of 1–1000 ppm.

Several fluorescence oil-in-water monitors that require calibration with an oil of the type to be monitored are commercially available.[122,123]

A recent portable instrument called Field Identification Luminescence Monitor (FILM) developed for Coast Guard use by Brownrigg et al.[124] employs a low-pressure Hg lamp as a source, a polychromator, and photographic detection to classify oil samples in the field. A unique feature of this instrument is its use of a density step filter, which results in a density step function proportional to the fluorescence intensity at each wavelength on the photographic print. The photographs show structural features roughly similar to the normal fluorescence emission spectra as produced by a conventional spectrofluorometer. This instrument has some oil screening capability for preliminary oil identification and may be used either on oil films or on oil dissolved in an organic solvent or water.

G. CONCLUSIONS

The general utility of fluorescence for the detection and identification of weathered oils has been well established. In this respect, fluorescence has become a mature technique and requires the solution of such related problems as improved correction for weathering effects and the use of mathematical pattern recognition techniques for more quantitatively exact matching schemes for oils.

Several promising fluorescence/luminescence techniques such as derivative and contour spectroscopy have not yet been fully exploited, especially for weathered oils. Low-temperature luminescence, although generally shown to produce more structured spectra, has also not been widely used.

Fluorescence quantitation, although used as the basis for several commercially available oil detectors and oil-in-water monitors, has not been developed as a standard analytical procedure. The research areas that require the greatest developmental efforts are quantitation, on-site monitoring, and remote sensing. Progress in the United States in the area of remote sensing of petroleum oils by fluorescence with laser excitation has been relatively slow compared to development elsewhere.

It is important that fluorescence spectroscopy should not rely on past accomplishments in the area of forensic oil spill identification. A continued effort is needed to keep up with state-of-the-art techniques and to solve remaining problems in the identification of heavily weathered oils.

REFERENCES

1. C. A. Parker, *Photoluminescence of Solutions* (Elsevier, New York, 1968).
2. D. Eastwood, S. H. Fortier, and M. S. Hendrick, *Am. Lab.* **10**(3), 45 (1978).
3. A. P. Bentz, *Anal. Chem.* **48**, 454A (1976).
4. D. Eastwood and M. S. Hendrick, paper presented at the Forensic Analysis Symposium at Eastern Analytical Symposium, New York, 1979.
5. T. A. Kubic, T. Kanabrocki, and J. Dwyer, paper presented at the American Academy of Forensic Sciences, 32nd Annual Meeting, 1980.
6. S. A. Wise, S. N. Chesler, H. S. Hertz, L. R. Hilpert, and W. E. May, *Carcinog. Compr. Surv.* **3**, 175 (1978).
7. U. Frank, *Toxicol. Environ. Chem. Rev.* **2**, 163 (1978).
8. W. A. Coakley, *Proceedings, Prevention and Control of Oil Spills*, Washington, D.C. (1973).
9. E. R. Adlard, *J. Inst. Pet.* (*London*) **58**, 63 (1972).
10. C. S. Woo, A. P. D'Silva, and V. A. Fassel, *Anal. Chem.* **52**, 159 (1980).
11. R. J. Hurtubise, J. F. Scharbon, J. D. Feaster, D. H. Therkildsen, and R. E. Poulson, *Anal. Chim. Acta*, **89**, 377 (1977).
12. U. Frank, D. Stainken, and M. Gruenfeld, *Oil Spill Conference* (1979), p. 323.
13. J. F. McKay and D. R. Latham, *Prepr. Pap. Nat. Meet. Am. Chem. Soc.* **16**, 63 (1972).
14. A. Ambruso, *Abstracts of Pittsburgh Conference*, Cleveland, Ohio (1974).
15. J. R. Jadamec and T. J. Porro, *Abstracts of Pittsburgh Conference*, Cleveland, Ohio (1974).
16. "Classification of Oil Products by Fluorometric Techniques," progress report prepared for U.S. Coast Guard R & D Center, Avery Point, Groton, Connecticut, by Beckman Instruments, Contract No. DOT-CG-33184-A (July 1974).
17. F. P. Schwarz and S. Wasik, *Abstracts of Pittsburgh Conference*, Cleveland, Ohio (1976).
18. J. R. Jadamec, J. E. Sheridan, T. J. Porro, and D. A. Terhaar, "Fingerprinting of Petroleum Oils of Fluorescence Spectroscopy," Pittsburgh Conference, No. 219, Cleveland, Ohio, 1974.
19. *ASTM Book of Standards* (1978), p. 720, D3650–78.

20. "Oil Spill Identification System," Chemistry Branch, U.S. Coast Guard R & D Center, Report No. DOT CG D-52-77 (June 1977).

21. J. B. F. Lloyd, *J. Forens. Sci. Soc.* **11**, 83 (1971).

22. J. B. F. Lloyd, *J. Forens. Sci. Soc.* **11**, 235 (1971).

23. J. B. F. Lloyd, *J. Forens. Sci. Soc.* **11**, 153 (1971).

24. J. B. F. Lloyd, *Analyst* **99**, 729 (1974).

25. T. Vo-Dinh, R. B. Gammage, and A. R. Hawthorne, *Polynuclear Aromatic Hydrocarbons, 3rd International Symposium of Chemical Biology—Carcinogens and Mutagens* (1978), p. 111.

26. T. Vo-Dinh, R. B. Gammage, and A. R. Hawthorne, *Abstracts of Pittsburgh Conference*, Cleveland, Ohio (1979).

27. J. R. Jadamec, S. Fortier, S. Buchanan, and G. L. Hufford, presentation at the Society of Naval Architects and Marine Engineers, New London, Connecticut, 1978.

28. S. G. Wakeham, *Environ. Sci. Technol.* **11**, 272 (1977).

29. U. Frank and M. Gruenfield, *Anal. Qual. Control Newsl. (U.S. EPA)* **32**, (1977).

30. P. John and I. Soutar, *Anal. Chem.* **48**, 520 (1976).

31. P. John and I. Soutar, *Proc. Anal. Div. Chem. Soc.* **13**, 309 (1976).

32. T. A. Kubic, C. M. Kanabrocki, and J. Dwyer, *"The Introduction of Variable Separation Synchronous Excitation Fluorescence and Its Application in Forensic Oil Identification,"* unpublished.

33. C. M. Kanabrocki, "The Introduction of Variable Separation Synchronous Excitation Fluorescence and Its Application in Forensic Oil Identification," B.S. independent research, C.W. Post College, New York, 1979.

34. T. J. Porro and D. A. Terhaar, *Anal. Chem.* **48**, 1103A (1976).

35. J. E. Sheridan and J. R. Jadamec, *Abstracts of Pittsburgh Conference*, Cleveland, Ohio (1976).

36. T. C. O'Haver, in *Modern Fluorescence Spectroscopy*, Vol. 1, E. L. Wehry, Vol. 1 (Plenum Press, New York, 1976).

37. G. L. Green and T. C. O'Haver, *Anal. Chem.* **46**, 2191 (1974).

38. J. L. DiCesare and T. J. Porro, *Trends Fluorom.* **1**, 16 (1978).

39. T. E. Cook, R. E. Santini, and H. L. Pardue, *Anal. Chem.* **49**, 871 (1977).

40. D. A. Kolb and K. K. Shearin, *Abstracts of Pittsburgh Conference*, Cleveland, Ohio (1977).

41. A. W. Hornig, *Proceedings, Pattern Recognition Applied to Oil Identification*, Coronado, California (1976).

42. A. W. Hornig and J. T. Brownrigg, *Abstracts of Pittsburgh Conference*, Cleveland, Ohio (1975).

43. A. W. Hornig and H. J. Coleman, *Abstracts of Pittsburgh Conference*, Cleveland, Ohio (1974).

44. I. M. Warner, J. B. Callis, E. R. Davidson, and G. D. Christian, *Proceedings, Pattern Recognition Applied to Oil Identification*, Coronado, California (1976).

45. I. M. Warner, G. D. Christian, E. R. Davidson, and J. B. Callis, *Anal. Chem.* **49**, 564, (1977).

46. L. P. Giering, *Ind. Res. Dev.*, 134 (1978).

47. L. P. Giering and A. W. Hornig, *Am. Lab.* **9**(11), 113 (1977).

48. J. T. Brownigg and A. W. Hornig, "Low Temperature Total Luminescence Contour Spectra of Six Topped Crude Oils and Their Vacuum Distillate and Residuum Fractions," U.S. Energy Research and Development Administration Bartlesville, Energy Technology Center, Bartlesville, Oklahoma; Order No. EY-77-X-19-2570(P) (1978).

49. J. T. Brownrigg and A. W. Hornig, *Abstracts of Pittsburgh Conference*, Cleveland, Ohio (1977).
50. L. P. Giering and A. W. Hornig, *Abstracts of Pittsburgh Conference*, Cleveland, Ohio (1977).
51. B. R. Chisholm, H. G. Eldering, L. P. Giering, and A. W. Hornig, "Total Luminescence Contour Spectra of Six Topped Crude Oils," Energy Research and Development Administration Bartlesville, Oklahoma, Order No. BE-76-P-1221 (1976).
52. G. Sogliero, D. Eastwood, L. Giering, and B. Chisholm, private communication, 1980.
53. S. H. Fortier and D. Eastwood, *Abstracts of Pittsburgh Conference*, Cleveland, Ohio (1976).
54. A. W. Hornig, D. Eastwood, J. Guilfoyle, and F. Kawahara, paper presented at the Pacific Conference on Chemical Spectroscopy, San Francisco, California, 1972.
55. A. W. Hornig, D. Eastwood, and J. Guilfoyle, "Development of a Low Temperature Molecular Emission Method for Oils," Program No. 16020 GBW, prepared for EPA Water Quality Office by Baird-Atomic, Inc. (August 1971).
56. A. W. Hornig and D. Eastwood, "Development of a Low Temperature Molecular Emission Method for Oils," Program No. 16202 GBW, prepared for EPA Water Quality Office by Baird-Atomic, Inc. (October 1971).
57. S. H. Fortier and D. Eastwood, *Anal. Chem.* **50**, 334 (1978).
58. D. Eastwood and M. S. Hendrick, *Abstracts of Pittsburgh Conference*, Cleveland, Ohio (1978).
59. D. Eastwood and M. Gouterman, unpublished results, 1969.
60. D. Eastwood and J. J. Hanks, *Abstract of Pittsburgh Conference*, Cleveland, Ohio (1976).
61. J. J. Hanks, "Oil Fingerprinting by Phosphorescence—A Feasibility Study," Academy Scholar's Report—USCG Academy, 1975.
62. R. I. Personov, *Abstracts of 21st Colloquium Spectroscopicum Internationale and 9th International Conference Atomic Spectroscopy, Cambridge, United Kingdom* (1979).
63. B. S. Causey, G. F. Kirkbright, and C. G. DeLima, *Analyst* **101**, 367 (1976).
64. R. Farooq and G. F. Kirkbright, *Analyst* **101**, 566 (1976).
65. G. F. Kirkbright and C. G. DeLima, *Analyst* **99**, 338 (1974).
66. R. B. Dickinson, Jr., R. R. Gore, and E. L. Wehry, *Abstracts of Pittsburgh Conference*, Cleveland, Ohio (1978).
67. E. L. Wehry, G. Mamantov, R. R. Kemmerer, E. R. Hinton, R. C. Stroupe and G. Goldstein, *Abstracts of Pittsburgh Conference*, Cleveland, Ohio (1979).
68. R. C. Stroupe, P. Tokousbalides, R. B. Dickinson, Jr., E. L. Wehry, and G. Mamantov, *Anal. Chem.* **49**, 700 (1977).
69. E. L. Wehry, *Fluoresc. News* **8**, 21 (1974).
70. J. F. Fantasia, T. M. Hard, and H. C. Ingrao, "An Investigation of Oil Fluorescence as a Technique for the Remote Sensing of Oil Spills," DOT Project No. 714104/A/003 (June 1971).
71. R. Fugate, private communication, 1980.
72. H. V. Drushel and A. L. Sommers, *Anal. Chem.* **38**, 10 (1966).
73. J. F. McKay and D. R. Latham, *Anal. Chem.* **45**, 1050 (1973).
74. J. R. Jadamec, W. A. Saner, and Y. Talmi, *Anal. Chem.* **49**, 1316 (1977).
75. W. A. Saner, J. R. Jadamec, K. Kallett, S. Cravitt, and D. Baker, *Abstracts of Pittsburgh Conference, Cleveland, Ohio* (1977).
76. Y. Talmi, D. C. Baker, J. R. Jadamec, and W. A. Saner, *Anal. Chem.* **50**, 936A (1978).
77. J. T. Brownrigg and A. W. Hornig, *Abstracts of Pittsburgh Conference*, Cleveland, Ohio (1976).
78. C. N. Ho, G. D. Christian, and E. R. Davidson, *Anal. Chem.* **50**, 1108 (1978).

79. J. R. Jadamec, W. A. Saner, and E. Kallet, *Abstracts of Pittsburgh Conference*, Atlantic City, New Jersey (1980).
80. A. M. Hornig and L. P. Giering, *Abstract of Pittsburgh Conference*, Cleveland, Ohio (1978).
81. D. C. Gordon, P. D. Keizer, W. R. Hardstaff, and D. C. Aldous, *Environ. Sci. Technol.* **10**, 580 (1976).
82. J. W. Frankenfeld, "Weathering of Oil at Sea," Report No. DOT-CG-D-7-75 (1975).
83. D. B. Boylan and B. W. Tripp, *Nature* **130**, 44 (1971).
84. D. Eastwood, M. S. Hendrick, and S. H. Fortier, *Abstracts of Pittsburgh Conference*, Cleveland, Ohio (1977).
85. T. J. Killeen, D. Eastwood, and M. S. Hendrick, *Abstracts of Pittsburgh Conference*, Cleveland, Ohio (1978).
86. M. L. Curtis, "Use of Pattern Recognition for Typing and Identification of Oil Spills," U.S. Coast Guard Research and Development Center, Avery Point, Groton, Connecticut, Report No. DOT-CG-D-38-77 (1977).
87. S. A. Starks and M. L. Curtis, *Proceedings, Pattern Recognition Applied to Oil Identification*, Coronado, California (1976).
88. Y. T. Chien and T. J. Killeen, *Proceedings, Pattern Recognition Applied to Oil Identification*, Coronado, California (1976).
89. Y. T. Chien and T. J. Killeen, *Abstracts of Pittsburgh Conference*, Cleveland, Ohio (1977).
90. T. J. Killeen and Y. T. Chien, *Proceedings, Pattern Recognition Applied to Oil Identification*, Coronado, California (1976).
91. T. J. Killeen, D. Eastwood, and M. S. Hendrick, *Talanta* **28**, 1 (1981).
92. L. P. Giering and A. W. Hornig, *Abstracts of Pittsburgh Conference*, Cleveland, Ohio (1978).
93. D. C. Gordon and P. D. Keizer, "Estimation of Petroleum Hydrocarbons in Sea Water by Fluorescence Spectroscopy: Improved Sampling and Analytical Methods," Technical Report 481, Fisheries Research Board of Canada (1974).
94. P. D. Keizer, D. C. Gordon, Jr., and J. Dale, *J. Fish. Res. Board of Can.* **34**, 347 (1977).
95. D. C. Gordon, P. D. Keizer, and J. Dale, *Mar. Chem.* **2**, 251 (1974).
96. P. D. Keizer and D. C. Gordon, *J. Fish. Res. Board Can.* **30**, 1039 (1973).
97. E. J. Hoffman, J. G. Quinn, J. R. Jadamec, and S. H. Fortier, *Bull. Environ. Contam. Toxicol.* **23**, 536 (1979).
98. R. E. Grojean, J. A. Sousa, J. F. Roach, E. F. Wyner, and M. Nakashima, *Opt. Eng.* **17**, 139 (1978).
99. A. P. Bentz, *Adv. Chem. Ser.* **185**, 55 (1980).
100. J. C. Mourlon, *Twelfth International Symposium on Remote Sensing of Environment* (1978).
101. H. G. Gross and M. Muramoto, *Remote Sensing at Earth Resources*, Vol. 3 (1974).
102. T. Sato, Y. Suziki, H. Kashiwagi, M. Nanjo, and Y. Kakui, *IEEE J. Ocean. Eng.* **3**, 1 (1978).
103. D. M. Rayner and R. O'Neil, *Opt. News* **5**(3), 13 (1979).
104. R. M. Measures, W. R. Houston, and D. G. Stephenson, *Opt. Eng.* **13**, 494 (1974).
105. R. M. Measures, J. Garlick, W. R. Houston, and D. G. Stephenson, *Can. J. Remote Sens.* 95 (1975).
106. R. M. Measures, *Proceedings of Optical Society of America Topical Meeting on Applications of Laser Spectroscopy*, Anaheim, California (1975).
107. G. A. Sandness and S. B. Ailes, *Proceedings of the Eleventh International Symposium on Remote Sensing of Environment*, Vol. 2, 1445 (1977).

108. A. W. Hornig, "Model Oil Study," Contract No. 68–01–0146, prepared for EPA Water Quality Office by Baird-Atomic, Inc. (December 1971).

109. A. W. Hornig and D. Eastwood, "A Study of Marine Luminescence Signatures," NASA Cr. 11457, Contract No. NAS2-6408 by Baird-Atomic Inc., Bedford, Massachusetts (1973).

110. H. G. Gross, *Proceedings of the Tenth International Symposium on Remote Sensing of Environment*, Vol. 2, 253 (1975).

111. R. D. Watson, W. R. Hemphill, and R. C. Bigelow, *Proceedings of the Tenth International Symposium on Remote Sensing of Environment*, Vol. 1, 203 (1975).

112. H. G. Eldering, A. W. Hornig, and W. A. Webb, "Detection and Identification of Oil Spills by Remote Fluorometric Systems," Baird-Atomic, Inc., Report No. DOT-CG-D-73-75 (September 1974).

113. H. R. Gram, "Construction, Installation, and Testing of an Oil Detection Buoy System and Related Effort on the Rouge River in Detroit, Michigan," Contract No. DOT-CG-63268-A (1978).

114. H. R. Gram, "Modification and Field Evaluation of the Oil Detection Buoy System," Contract No. DOT-CG-53356-A (1976).

115. H. R. Gram, "Petroleum Oil Detection Buoy System," Final Report No. DOT-CG-D-39-75 (1975).

116. H. R. Gram, *Abstracts of Pittsburgh Conference*, Cleveland, Ohio (1975).

117. H. R. Gram, *Abstracts of Pittsburgh Conference*, Cleveland, Ohio (1974).

118. A. W. Hornig, J. T. Brownrigg, and B. R. Chisholm, "A Shipboard Oil-in-Water Content Monitor Based on Oil Fluorescence," Final Report DOT-CG-D-54-76 (1976).

119. A. W. Hornig, J. T. Brownrigg, B. R. Chisholm, and L. P. Giering, *Proceedings, Oil Spill Conference* (1977).

120. A. W. Hornig and J. T. Brownrigg, "A Shipboard Oil-in-Water Content Monitor based on Oil Fluorescence," Baird-Atomic, Inc., Report No. DOT-CG-D-87-75 (February 1975).

121. A. W. Hornig and B. R. Chisholm, *SPIE Unconv. Spectrosc.* **82**, 97 (1976).

122. Baird-Petroleum Oil in Water Monitor, commercially available from Baird Co., Bedford, Massachusetts.

123. Clarke Chapman Standard Oil Pollution Monitor Type 4967, commercially available from Clarke Chapman Ltd., United Kingdom.

124. J. T. Brownrigg, A. W. Kliman, and J. R. Jadamec, *Abstracts of Pittsburgh Conference*, Atlantic City, New Jersey (1980).

Index